SpringerBriefs in Environmental Science

SpringerBriefs in Environmental Science present concise summaries of cutting-edge research and practical applications across a wide spectrum of environmental fields, with fast turnaround time to publication. Featuring compact volumes of 50 to 125 pages, the series covers a range of content from professional to academic. Monographs of new material are considered for the SpringerBriefs in Environmental Science series.

Typical topics might include: a timely report of state-of-the-art analytical techniques, a bridge between new research results, as published in journal articles and a contextual literature review, a snapshot of a hot or emerging topic, an in-depth case study or technical example, a presentation of core concepts that students must understand in order to make independent contributions, best practices or protocols to be followed, a series of short case studies/debates highlighting a specific angle.

SpringerBriefs in Environmental Science allow authors to present their ideas and readers to absorb them with minimal time investment. Both solicited and unsolicited manuscripts are considered for publication.

Angela Schlutow • Winfried Schröder

Climate Change and Atmospheric Deposition as Drivers of Forest Ecosystem Integrity and Services

A Methodology for Assessing and Mapping Ecosystem Services across Time and Space

 Springer

Angela Schlutow
Neuenhagen, Germany

Winfried Schröder
Chair of Landscape Ecology
University of Vechta
Vechta, Germany

ISSN 2191-5547   ISSN 2191-5555 (electronic)
SpringerBriefs in Environmental Science
ISBN 978-3-031-67102-9   ISBN 978-3-031-67103-6 (eBook)
https://doi.org/10.1007/978-3-031-67103-6

This Springer imprint is published by the registered company Springer Nature Switzerland AG
The registered company address is: Gewerbestrasse 11, 6330 Cham, Switzerland

If disposing of this product, please recycle the paper.

Contents

Abbreviations

Al	Alluvial soils in broad river valleys, including terraces and lowlands
BArtSchV	Federal Species Protection Ordinance
BERN	Model and database "Bioindication for the regeneration of ecosystems to natural conditions"
BÜK1000N	Reference soil profile of the land use-specific soil map 1:1,000,000 Germany
°C	Degree Celsius
C/N	Carbon/nitrogen ratio
CICES	Common classification of ecosystem services
CLC	Corine Landcover (UBA 2015)
CO_2	Carbon dioxide
C_{org}	Organic carbon
D	Diluvial soils in the hilly lowlands and hill country
DWD	German Weather Service
EC REGULATION	Annex B VO 1332/2005 EC species protection regulation
EEA	European Environment Agency
EMEP	European cooperation programme for the monitoring and assessment of long-range transmission of air pollutants in Europe
eq	Equivalent of hydrogen ions
EU	European Union
FFH DIRECTIVE	Fauna-Flora-Habitat Directive
FK	Field capacity
Hh	Raised bog, fen
K	Soils in the coastal regions
Delete	Soils in the loess areas
Ls2-4	Weak to strong sandy loam
Lt2	Weak clayey loam
Lt3	Medium clayey loam
Lts	Sandy-loamy loam

Lu	Silty loam
MAES	Mapping and assessment of ecosystems and their services (MAES 2016)
N	Nitrogen
NH_3_dep	Atmospheric deposition rate of ammonia [eq m^{-2} a^{-1}]
NO_x_dep	Atmospheric deposition rate of oxidised nitrogen compounds [eq m^{-2} a^{-1}]
$pH_{(H2O)}$ value	Value of the negative decadic logarithm of the hydrogen concentration measured in water
RCP	Representative concentration paths
S_dep	Atmospheric deposition rate of sulphur compounds [eq m^{-2} a^{-1}]
Sl2	Light, loamy sand
Sl3	Medium loamy sand
Sl4	Very loamy sand
Slu	Silty-loamy sand
Ss	Reiner Sand
St2	Slightly loamy sand
St3	Medium loamy sand
STAR	Statistical regional climate model
Su2	Light silty sand
Su3	Medium silty sand
Su4	Highly silty sand
Ts2	Slightly sandy clay
Ts3	Medium sandy clay
Ts4	Very sandy clay
Tt	Pure clay
Tu2-4	Weak to strong silty clay
Uls	Sandy-clayey silt
We	Sandy silt
Ut2-4	Weak to strong clayey silt
Uu	Pure silt
V	Weathered soils from solid rock and the surrounding rock masses in mountainous and hilly landscapes
Vg	Rock-rich weathered soils of the high mountains
VSD+	Very Simple Dynamics Soil model, version 5.3.1 (Bonten et al. 2016; CCE 2012; Posch and Reinds 2009)
WMO	World Meteorological Organisation

Chapter 1
Introduction

Abstract The condition of ecosystems and their services are influenced in particu-lar by climate change and atmospheric nitrogen deposition. Due to this, the European Union has set the goal of assessing and mapping the condition of ecosystems and ecosystem services at the Union and Member State level in order to implement conservation or protection measures where necessary. So far, there is no rule-based quantitative methodology for this. Therefore, this contribution presents a methodol-ogy with which ecosystem services can be ordinally classified and mapped in a rule-based, transparent and automated way on the basis of empirical data and data modelled for projections for four spatial scales by examples of Germany.

Keywords Ecosystem services · CICES · Matrix method · Rule-based assessment · Ordinal classification

Ecosystem services are the benefits of ecosystems for human society (Ehrlich and Ehrlich 1981; Herrmann et al. 2011; MEA 2005). They depend on ecosystem struc-tures (e.g., biotic and abiotic ecosystem elements, "resources") and on their ener-getic and material relationships, i.e., on their functions based on biological, chemical and physical processes (Ruskule et al. 2018). According to Target 2 Action 5 of the European Biodiversity Strategy, all EU Member States are obliged to "map and assess the state of ecosystems and their services within their territory" (EU 2011, p. 5).

The Common International Classification of Ecosystem Services (CICES) has evolved from the work of the European Environment Agency on environmental accounting (EEA 2023). The aim of the common international classification is to standardise the description of ecosystem services. The standardisation of environ-mental accounting, work on the mapping and assessment of ecosystem services and the assessment of ecosystems in general would benefit from more systematic approaches to naming and describing ecosystem services.

The first fully functional version of CICES (V4.3) was released in 2013. Based on the experience gained by the user community since then, the structure and scope

A. Schlutow, W. Schröder, *Climate Change and Atmospheric Deposition as Drivers of Forest Ecosystem Integrity and Services*, SpringerBriefs in Environmental Science, https://doi.org/10.1007/978-3-031-67103-6_1

of the system have been revised and a fully revised version (V5.1) has been available since 2023 (EEA 2023).

Methodological guidelines have been developed to support the individual EU Member States in implementing this measure (Maes et al. 2013, 2016). A common method for classifying ecosystem services is the so-called matrix method (Jacobs et al. 2015). In this method, ecosystem services are categorised on a scale from 0 "insignificant" to 5 "very high" by expert assessment and linked to various spatial mapping units (Burkhard and Maes 2017; Maes et al. 2016). The EU recommends the method for spatial visualisation and "rapid assessment" (Jacobs et al. 2015) of ecosystem services as part of the European Biodiversity Strategy (Science for Environment Policy EU 2015).

One methodological problem with the matrix method is that the assessment is not always based on quantitative information and is not rule-based. There is often a lack of methodological transparency as a prerequisite for objective, reliable (reproducible) and valid results (Schröder and Nickel 2020; Schlutow and Schröder 2021). In addition, the publications on the application of the matrix method lack information on the variability of the experts' assessments and thus on the objectivity of the method as well as on whether repeated assessments by the same experts allow an assessment of the reproducibility of the expert assessments. Sohel et al. (2015) therefore emphasise the methodologically necessary consideration of quantitative biophysical indicators and empirical modelling.

Schröder et al. (2020) presented a methodological approach for the classification of current forest ecosystem types ("ANOEST") in Germany, which makes the changes in ecosystem integrity accessible by comparing prospective ecosystem states with ecosystem type-specific reference states, as described by quantitative indicators, for 61 forest ecosystem types based on data from 1960 to 1990. The input of acidifying and eutrophying air pollutants into German ecosystems peaked between 1960 and 1990 and climate change had already begun. The approach presented here is intended to supplement the ANOEST approach with the definition of more or less pristine states from the time before the industrialisation wave in the second half of the twentieth century. The aim of this paper was to develop and present a methodology for the rule-based, transparent and automated ordinal classification and mapping of ecosystem services based on monitoring data from the period before the industrial revolution and modelled data for projections of potential future ecosystem states and associated services for different spatial scales in Germany. The in-depth rule-based assessment and mapping of ecosystem services focussed on three services, as they are of particular importance according to the Common International Classification of Ecosystem Services (Haines-Young and Potschin 2018): habitat services (Sect. 2.2), carbon sequestration (Sect. 2.4) and the service of primary biomass production (Sect. 2.3). An additional rule-based assessment of other ecosystem services included in CICES according to Haines-Young and Potschin (2018) was also carried out, but only at a rough level (Sect. 2.5). This is necessary in order to check not only the stratification of abiotic and biotic parameters, but also to ensure that as many ecosystem service potentials as possible match when summarising ecosystem types into ecosystem type classes (see Sect. 3.1.1).

In addition to a variety of environmental factors, climate change and atmospheric nitrogen deposition can alter the integrity of ecosystems, i.e., their structures and functions, and thus also limit their benefits to humans, i.e., ecosystem services (Schlutow and Schröder 2021). Therefore, the methodology to be developed should make it possible to assess whether ecosystem structures and functions are moving significantly away from a pristine state and to identify stages of change in ecosystem integrity up to the replacement of one ecosystem type by another using quantitative data from current environmental monitoring and modelling. The desired methodology should allow to ordinally classify (ordinate, grade) changes in ecosystem integrity at different spatial scales. To achieve this, the methodology relied on an extensive vegetation database and on nationally available up-to-date data from monitoring and mapping programmes. It was supplemented by dynamic modelling of possible future ecosystem conditions under climate change and atmospheric nitrogen deposition (Schlutow and Schröder 2021). With regard to the modelling of ecosystem services, the methodology presented can be described as a "Tier 3 approach" (Burkhard and Maes 2017; Ruskule et al. 2018).

References

Burkhard B, Maes J (eds) (2017) Mapping ecosystem services. Pensoft Publishers, Sofia, p 376

EEA (European Environmental Agency) (2023) CICES Version 5.1. https://cices.eu/

Ehrlich P, Ehrlich A (1981) Extinction: causes and consequences of the disappearence of species. Random House, New York

EU (European Union) (2011) The EU biodiversity strategy to 2020. Publications Office of the European Union, Luxembourg, ISBN 978-92-79-20762-4. https://doi.org/10.2779/39229

Haines-Young R, Potschin MB (2018) Common international classification of ecosystem services (CICES) V5.1 and guidance on the application of the revised. Structure. Available from www.cices.eu

Herrmann A, Schleifer S, Wrbka T (2011) The concept of ecosystem services regarding landscape research: a review. Living Rev Landsc Res 5:1. http://www.livingreviews.org/lrlr-2011-1

Jacobs S, Burkhard B, Van Daele T, Staes J, Schneiders A (2015) 'The matrix reloaded': a review of expert knowledge use for mapping ecosystem services. Ecol Model 295:21–30

Maes J, Teller A, Erhard M, Liquete C, Braat L, Berry P, Egoh B, Puydarrieux P, Fiorina C, Santos F, Paracchini ML, Keune H, Wittmer H, Hauck J, Fiala I, Verburg PH, Condé S, Schägner JP, San Miguel J, Estreguil C, Ostermann O, Barredo JI, Pereira HM, Stott A, Laporte V, Meiner A, Olah B, Royo Gelabert E, Spyropoulou R, Petersen JE, Maguire C, Zal N, Achilleos E, Rubin A, Ledoux L, Brown C, Raes C, Jacobs S, Vandewalle M, Connor D, Bidoglio G (2013) Mapping and assessment of ecosystems and their services. An analytical framework for ecosystem assessments under Action 5 of the EU Biodiversity Strategy to 2020, Discussion paper—final, April 2013. Publications office of the European Union, Luxembourg

Maes J, Liquete C, Teller A, Erhard M, Paracchini ML, Barredo JI, Grizzetti B, Cardoso A, Som-ma F, Petersen J-E, Meiner A, Gelabert ER, Zal N, Kristensen P, Bastrup-Birk A, Biala K, Piroddi C, Egoh B, Degeorges P, Fiorina C, Santos-Martín F, Naruševičius V, Verboven J, Pereira HM, Bengtsson J, Gocheva K, Marta-Pedroso C, Snäll T, Estreguil C, San-Miguel-Ayanz J, Pérez-Soba M, Grêt-Regamey A, Lilleb AI, Malak DA, Cond S, Moen J, Czúcz B, Drakou EG, Zulian G, Lavalle C (2016) An indicator framework for assessing ecosystem services in support of the EU biodiversity strategy to 2020. Ecosyst Serv 17:14–23

MEA (Millennium Ecosystem Assessment) (2005) Ecosystems and human well-being: synthesis. Island Press, Washington, DC

Ruskule A, Vinogradovs I, Villoslada PM (2018) The guidebook on ecosystem service framework and its application in integrated planning, Version 28.09.2018. University of Latvia, Faculty of Geography and Earth Sciences. Riga:1–63

Schlutow A, Schröder W (2021) Rule-based classification and mapping of ecosystem services with data on the integrity of forest ecosystems. Environ Sci Eur 33:50. https://doi.org/10.1186/s12302-021-00481-3

Schröder W, Nickel S (2020) Research data management as an integral part of the research process of empirical disciplines using landscape ecology as an example. Data Sci J 19(26):1–14. https://doi.org/10.5334/dsj-2020-026

Schröder W, Schlutow A, Dworcyk C, Jenssen M, Nickel S (2020) Regelbasierte Einstufung und Kartierung von Ökosystemleistungen mit Daten zur Integrität von Waldökosystemtypen. In book: Handbuch der Umweltwissenschaften. 28. Erg. Lfg. [Rule-based classification and mapping of ecosystem services with data on the integrity of forest ecosystem types. In book: Handbook of Environmental Sciences.]. https://www.researchgate.net/publication/339995889_Regelbasierte_Einstufung_und_Kartierung_von_Okosystemleistungen_mit_Daten_zur_Integritat_von_Waldokosystemtypen

Science for Environment Policy (2015) Ecosystem services and the environment. In-depth report 11 produced for the European Commission, DG Environment by the Science Communication Unit. UWE, Bristol. http://ec.europa.eu/science-environment-policy

Sohel MSI, Ahmed MS, Burkhard B (2015) Landscape's capacities to supply ecosystem services in Bangladesh. A mapping assessment for Lawachara National Park. Ecosyst Serv 12:128–135

Chapter 2
Rule-Based Methodology for Ordinating and Classification of Ecosystem Services

Abstract Three ecosystem services of ecosystems in the pristine condition (before 1960) and under the influence of land use, climate change and nitrogen pollution currently (1991–2010) and expected in the future (until 2070) were quantified for 78 forest ecosystem type classes in Germany: habitat service, carbon sequestration service and primary biomass production. The respective ecosystem service classifications were mapped for the territory of Germany, for the federal state Saxony and for a Long Term Ecological Researched site (LTER). The rules for evaluating the three ecosystem services studied in depth were also applied to relevant open land biotopes.

Keywords Rules-based rating · Habitat service · Carbon sequestration · Biomass production

2.1 Framework

The presented method of rule-based grading (categorisation, rating) of ecosystem services is based on objectively measurable criteria that can be applied to all cartographic scales. The mapping differs only in the spatial resolution of the cartographic basis, the data.

The development of the rule-based ordinal classification and mapping of ecosystem services focussed on the following three services, as they are of particular importance for regulation and conservation services and have large-scale significance:

1. Habitat Service (CICES: "Self-regulation and self-organisation of ecosystems"),
2. Carbon sequestration (CICES "Contribution to global climate regulation"),
3. Primary biomass production (tree wood, CICES: "plant and animal raw materials").

A. Schlutow, W. Schröder, *Climate Change and Atmospheric Deposition as Drivers of Forest Ecosystem Integrity and Services*, SpringerBriefs in Environmental Science, https://doi.org/10.1007/978-3-031-67103-6_2

5

The rating scale comprises six levels: 0 = ecosystem service potential without significance; 1 = ecosystem service potential with low significance; 2 = ecosystem service potential with medium significance; 3 = ecosystem service potential with medium significance; 4 = ecosystem service potential with high significance, 5 = ecosystem service potential with very high significance.

2.2 Habitat Service

The following criteria are particularly important for categorising the potential services of an ecosystem as a habitat for plants and animals (Fröhlich and Sporbeck 2002, expanded and modified):

(a) Hemerobicity (degree of naturalness)
(b) Compositional completeness
(c) Habitat value for fauna
(d) Vulnerability/need for protection
(e) Restorability/recoverability/replaceability of habitats
(f) Maturity
(g) Position within the biotope network

These criteria were then defined as follows and assigned to the six ecosystem service levels 0–5 (Sect. 2.1).

(a) Hemerobicity
Hemerobicity is the deviation of an ecosystem from the potentially natural type of vegetation due to antropogenic influences. Hemerobicity is therefore a measure of an ecosystem's ability to self-regulate. The rules for the assessment of Hemerobicity are contained in Table 2.1.

Table 2.1 Criteria, parameters and assessments for the ordinal classification of the degree of naturalness

Parameters	Scores
Natural—largely uninfluenced (pristine forests, natural unutilised open land)	5
Near-natural—anthropogenic changes recognisable (afforestation with native species)	5
Near-natural—deforested but little changed site (permanently utilised grassland)	4
Near-natural/semi-natural with significant impairments	3
Far away from nature	2
Far away from nature—with irreversibly changed location factors	1

A categorization of the degree of naturalness of plant communities (including antropogenic forest plantations, arable wild communities, intensive grassland, ruderal communities and hypertrophic aquatic plant communities) is included in the BERN database (Schlutow et al. 2024).

(b) Compositional completeness

The "National Strategy on Biological Diversity" (BMU 2007) serves to preserve and develop natural and near-natural forest communities. To this end, the degree of similarity between the species composition and vegetation structure of an ecosystem and its pristine state is used as a criterion for categorising the biodiversity of an ecosystem. The species combinations in the pristine state of the ecosystem type, which were quantitatively described at sites with little or no pollution around 1960 or before, serve as a reference. The comparison of these potential with the current species composition and its vegetation structure is the measure for the assessment of biodiversity in the sense of the above-mentioned biodiversity strategy. The rules for the assessment of the relative similarity of the vegetation as a measure of the deviation of the current ecosystem from the natural vegetation can be found in Table 2.2.

(c) Habitat value for fauna

The grading of the potential ecosystem service of an ecosystem as a habitat for animals is based on species groups that have more or less pronounced preferences for certain ecosystems as (partial) habitats. A list of animals with high significance in Germany and their preferred habitats is given by Schlutow and Schröder (2021—Supplement S1). A high score applies to ecosystems with as many (partial) habitat services as possible for as many animal groups as possible. Natural wet forests have

Table 2.2 Criteria, parameters and scores for the grading of vegetation biodiversity

Parameters	Scores
> 90–100% of the diagnostic plant species in the ecosystem	5
> 80–90% of the diagnostic plant species in the ecosystem	4
> 50–80% of the diagnostic plant species in the ecosystem	3
> 30–50% of the diagnostic plant species in the ecosystem	2
> 10–30% of the diagnostic plant species in the ecosystem	1
< 10% of the diagnostic plant species in the ecosystem	0

a very high habitat value for the indicator animal species. The anhydromorphic semi-natural forests and the semi-natural open land ecosystem have high habitat values. Managed forests generally have an average habitat value for their indicator species.

The rules for categorising suitability for faunal diversity as a measure of the deviation of the number of animals in a current ecosystem from the natural ecosystem can be found in Table 2.3.

(d) Vulnerability/need for protection

Ecosystems designated as flora-fauna-habitat types (Annex I Habitats Directive 2009/147/EC European Parliament 2009) are of particular importance as they provide habitats for protected species. The following regulations and directives identify habitats that require special protection (Table 2.4):

- Directive 2009/147/EC of the European Parliament (2009) and of the Council of 30 November 2009 on the conservation of wild birds Council Directive 92/43/EEC of 21 May 1992 on the conservation of natural habitats and of wild fauna and flora (Habitats Directive Annexes II and IV)
- Annex A Regulation 1332/2005 EC Species Protection Regulation (EC Regulation 2005)

Table 2.3 Criteria, parameters and scores for ordinating faunal diversity

Parameters	Scores
The ecosystem is also a breeding, feeding and refuge area for mammals, birds, reptiles, amphibians and insects	5
Ecosystem is also a place of reproduction, food and refuge for mammals, birds, reptiles, amphibians and insects.	4
An ecosystem is a breeding, feeding or refuge habitat for mammals, birds, reptiles and amphibians as well as insects.	3
An ecosystem is a breeding, feeding or refuge habitat for mammals, birds, reptiles, amphibians or insects.	2
An ecosystem is a breeding, feeding or refuge habitat for mammals, birds, reptiles, amphibians or insects.	0

Table 2.4 Criteria, parameters and scores for ordinally categorising the need for protection

Parameters	Scores
Ecosystem is FFH habitat type (Annex I FFH Directive)	5
The ecosystem is suitable for protected species under the Habitats Directive (Annex II and IV) and for specially and strictly protected bird species under the EU Birds Directive.	4
Ecosystem is suitable for occurrence of specially and strictly protected species	3
The ecosystem is suitable for the occurrence of a legally protected biotope.	2
Ecosystem is rare and endangered by utilisation pressure	1
The ecosystem is not a suitable location for the occurrence of protected/endangered species.	0

- Annex B Regulation 1332/2005 EC Species Protection Regulation (EC Regulation 2005),
- Federal Species Protection Ordinance (BArtSchV 2005) Annex 1, column 2 and
- Federal Species Protection Ordinance (BArtSchV 2005) Annex 1, column 3.

The assessment rules for the protection status of ecosystems as a measure of the need for protection are listed in Table 2.4.

(e) Restorability/recoverability/replaceability of habitats

If, following the planting/seeding/establishment of initial vegetation (trees, shrubs and/or dominant grass species, shaping of the soil and water balance), a functioning and self-regenerating ecosystem has developed within the time specified under "Restorability", the ecosystem is considered restored. The evaluation rules are summarised in Table 2.5.

(f) Maturity

Near-natural forests in the mature stage have the highest possible habitat potential for most forest-typical animal and plant species, as the native species are evolutionarily particularly well adapted to the climax forests that existed almost everywhere before the Middle Ages. The maturation of ecosystems is a non-linear process over time. A mature ecosystem is characterised by a mosaic of structures, uneven-aged mixed stands of different tree species and diverse compositions of the lower vegetation layers. Nevertheless, the term maturity is used here to describe a self-organising stage of dynamic forest development. The evaluation rules can be found in Table 2.6.

(g) Significance for the biotope network

The importance of an ecosystem in the biotope network was also measured by its function as a habitat or partial habitat for species with very large area requirements (e.g., raptors, large mammals, etc.). The categorisation of the habitat service of an ecosystem can be found in Table 2.7.

Table 2.5 Criteria, parameters and scores for ordinally categorising recoverability

Parameters	Scores
The ecosystem cannot be restored in the long term	5
The ecosystem is to be restored in the long term (in around 100–300 years).	4
The ecosystem is to be restored in the medium term (in approx. 60–100 years).	3
The ecosystem is to be restored in the medium term (in approx. 30–60 years).	2
The ecosystem must be restored in the medium term (in about 10–30 years).	1
Ecosystem to be restored in the short term (up to about 10 years).	0

Table 2.6 Criteria, parameters and scores for ordinally categorising the maturity level

Parameters	Scores
Peak reached	5
Peak achievable with the current vegetation (near-natural pre-climate stage).	4
Peak achievable with continued, non-anthropogenically disturbed ecosystem development (natural succession to potentially natural vegetation when utilisation is abandoned).	3
Climax cannot be achieved with the current vegetation due to the site conditions (non-native tree species).	2
High point not achievable with current vegetation due to land use (constantly disturbed sites).	1
Peak cannot be reached due to irreversible site deterioration	0

Table 2.7 Criteria, parameters and scores for ordinally categorising the function in the biotope network system

Parameters	Scores
Main networking element in the biotope network	5
Stepping stone biotope in the biotope network	3
Low importance in the biotope network	1
No significance for the biotope network	0

(h) Integrative assessment of ecosystem habitat services

When integrating the criteria-specific assessment (a)–(g), a ranking of the criteria was taken into account, which assumes that the individual criteria should not be included equally in an aggregation. The weighting was derived from the frequency with which the respective criterion was mentioned as a prerequisite for habitat services for the protected species relevant in Germany in Schlutow and Schröder (2021—Supplement S1). The criterion hemerobicity was ranked first and the following criteria were prioritised in the order in which they were mentioned. The following equation was then used to make additions to or subtractions from the hemerobia score:

$$L = N + \left((R-N)*0.5\right) + \left((H-N)*0.25\right) + \left((S-N)*0.125\right)$$
$$+ \left((W-N)*0.06\right) + \left((M-N)*0.03\right) + \left((B-N)*0.015\right),$$

where L is the score for overall habitat service; N is the hemerobia score for closeness to nature; R is the score for relative compositional completeness of flora and characteristic vegetation structure; H is the score for habitat value for fauna; S is the score for need for protection; W is the score for restorability; M is the score for maturity certificate; B is the score for position in the biotope network system.

2.3   Above-Ground Net Primary Biomass Production Service

The following criteria were of particular importance for ranking the service of the ecosystem type classes for primary biomass production:

(a) Plant physiological net primary productivity,
(b) specific soil fertility and
(c) Influence of climate on fertility.

The following method links the reference state of biomass productivity with the ecosystem type classes.

(a) Plant physiological net primary production
The ordinal classification of the plant species-specific net primary productivity of ecosystems must be based on the potential of the species at sites in a sustainable ecological balance of nutrients, water and energy (Balla et al. 2013). For this reason, yield tables and yield statistics collected at sites representing a more or less harmonious balance between the site factors and between these and the forest stands or representing this balance at the time of the respective survey, i.e., in particular surveys from the period before 1960, before high nitrogen inputs had led to disharmonious nutrient ratios (Schlutow and Scheuschner 2023), had to be analysed for the grading. The basis for the site-specific ordinal classification of the potential net primary productivity of forests were yield tables of tree species growth (Böckmann 1990; Jüttner 1955; Knapp 1973; Schober 1975; Schober 1987; Schwappach 1912; Wiedemann 1936; Wiedemann 1943; Wimmenauer 1919). The average annual increment for yield class 1 and the worst yield class of the tree species were determined from the yield tables over 100 years. The fixed growth rates determined in this way (DGZ 100) were converted into weight growth rates using the tree species-specific wood and bark density (Schlutow et al. 2018). It was assumed that the bark is removed from the stand, as is currently customary. The evaluation of the tree species-specific net primary productivity (Table 2.8) was carried out with evaluation points from 0 to 5, whereby 0 is not awarded, as every tree species and every soil has a net primary productivity.

(b) Determination of soil-specific net primary productivity
The method described below serves to concretise a discrete soil-specific value within the vegetation type-specific range of net primary productivity (Table 2.8), taking into account the different soil properties. This first requires the best possible estimation of soil fertility depending on the soil texture of the horizons of a rooted profile (Table 2.9).

The factors that influence the soil-specific yield potential (Table 2.9) were classified as shown in Table 2.10.

The classification rules in Table 2.10 were created on the basis of the following considerations and data:

Table 2.8 Intervals of net primary production (wood stage) of dominant and subdominant species in the virgin state

Tree species	Average annual growth rates after 100 years [DGZ 100].		Scores
	Net primary productivity of yield class I for logs with bark	Net primary productivity of the worst yield class for roundwood with bark	
	$E_{max(Phyto)}$ [t dry matter ha^{-1} a^{-1}]	$E_{min(Phyto)}$ [t dry matter ha^{-1} a^{-1}]	
Scots pine	3.5	1.4	3
Spruce	4.9	3.2	5
Mountain pine and black pine	0.8	0.7	1
Silver fir	3.5	3	3
European larch	2.5	2	2
Beech	4.9	2.4	5
Pedunculate and sessile oak	4	1.4	4
Alder	4.3	2.5	4
Birch, all species	2.8	2.1	2
Willows, all types	2.3	1.6	2
Elm, all species	4.8	3	4
Ash	3.7	2.5	4
Mountain ash	2.1	1.6	2
Lime, all species	5.2	3	5
Maple, all species	3.5	2.5	3
Poplars, aspen, all species	4.5	1.1	4

Sources: Böckmann (1990), Jüttner (1955), Knapp (1973), Schober (1975), Schober (1987), Schwappach (1912), Wiedemann (1936), Wiedemann (1943), Wimmenauer (1919)

Soil texture and pedogenesis:
The nomenclature of the soil texture classes was based on the German Soil Science Mapping Guidelines (AG Boden 2005). The soil texture classes were further subdivided according to pedogenesis (diluvial, alluvial, weathered soils).

Pores with dead water, plant-available adhesive water and air filled pores:
The volume fractions and diameters of the water- and air-filled pores as well as the suction tension of the different soil types were taken from Amberger (1988). The proportion of plant-available adhesive water (= usable field capacity) is highest in silt and sandy silt with an average of 26 vol.% and lowest in pure sands with approx. 10 vol.% for the different storage densities.

The proportion of pores filled with air that can be rooted through is highest in pure sands at 36% by volume and lowest in clays at 4% by volume. Optimum plant growth is achieved with a ratio of pores with available adhesive water to air-filled root penetration pores of 1:1 (Amberger 1988).

Table 2.9 Categorisation of soils according to pedogenesis and texture with regard to the influence on the potential yield of arable wood and grassland (Schlutow et al. 2021)

Soil condition according to the pedological mapping instructions 5th edition	Genesis	Water balance of the soil			Nutrient balance			Structure of the floor			Relative earnings potential EP$_{(geo\text{-}boz)}$
		Porosity < 0.2 µm with dead water (pF > 4.2)/ formation of waterlogging	Risk of dehydration	Groundwater or backwater influence	Utilisable field capacity (pore content 0.2–50 µm with plant-available adhesive water pF4.2–1.8)	Humus content depending on the soil composition	Cation exchange capacity	Thoroughness	Root penetration (pore content > 50 µm with air. pF<1.8)	Inclination of solidification in the B horizon	
Ss	D	5	1	1	1	1	1	3	5	1	2.00
	Al	5	1	4	1	1	1	4	5	1	2.36
	K	5	1	1	1	1	1	2	5	1	1.89
	V	5	1	1	1	1	1	2	5	1	1.89
Su2. Sl2. Sl3. St2	D	4	2	1	3	1	2	3	5	1	2.50
	Al	4	2	5	3	1	2	4	5	1	2.94
	K	4	2	4	3	1	2	3	5	1	2.75
	V	4	2	3	3	1	2	2	5	1	2.56
Su3. Su4	D	3	3	2	4	2	2	3	5	2	3.00
	lo	3	3	2	4	2	2	5	5	2	3.22
	Al.K	3	3	5	4	2	2	4	5	2	3.36
	V	3	3	3	4	2	2	2	5	2	2.97
	Vg	3	3	1	4	2	2	1	5	2	2.69

(continued)

Table 2.9 (continued)

Soil condition according to the pedological mapping instructions 5th edition	Genesis	Water balance of the soil			Nutrient balance			Structure of the floor			Relative earnings potential EP(geo-hoz)
		Porosity < 0.2 μm with dead water (pF > 4.2)/ formation of waterlogging	Risk of dehydration	Groundwater or backwater influence	Utilisable field capacity (pore content 0.2-50 μm with plant-available adhesive water pF4.2-1.8)	Humus content depending on the soil composition	Cation exchange capacity	Thoroughness	Root penetration (pore content > 50 μm with air. pF<1.8)	Inclination of solidification in the B horizon	
Slu. Sl4. St3	D	3	4	2	5	2	3	3	4	3	3.39
	lo	3	4	2	5	2	3	5	4	3	3.61
	Al.K	3	4	5	5	2	3	5	4	3	3.86
	V	3	4	3	5	2	3	2	4	3	3.36
	Vg	3	4	1	5	2	3	1	4	3	3.08
Ls2-4. Lt2. Lts. Ts4. Ts3	D	3	4	3	5	3	4	3	3	4	3.69
	lo	3	4	2	5	3	4	5	3	4	3.83
	Al	3	4	5	5	3	4	5	3	4	4.08
	K	3	4	4	5	3	4	3	3	4	3.78
	V	3	4	3	5	3	4	2	3	4	3.58
	Vg	3	4	1	5	3	4	1	3	4	3.31

Soil type group	Soil quality class										$EP_{(geo-hor)}$
Uu. Us. Ut2-4. Uls. Lu	D	2	3	3	4	4	3	3	2	5	3.33
	Al	2	5	5	4	4	4	5	2	5	4.00
	lo	2	3	2	4	4	4	5	2	5	3.58
	K	2	3	4	4	4	3	4	2	5	3.53
	V	2	3	3	4	4	3	2	2	5	3.22
	Vg	2	3	1	4	4	3	1	2	5	2.94
Lt3. Tu2-4. Ts2. Tl. tt	D	1	1	2	3	5	5	3	1	5	3.03
	Al	1	1	5	3	5	5	4	1	5	3.39
	lo	1	1	2	3	5	5	5	2	5	3.36
	K	1	1	4	3	5	5	1	1	5	2.97
	V	1	1	3	3	5	5	2	1	5	3.00
	Vg	1	1	2	3	5	5	1	1	5	2.81
Hh		1	1	1	1	5	1	1	1	1	1.44
Hn		1	1	1	2	5	3	4	3	1	2.42

$EP_{(geo-hor)}$ = Relative yield potential, determined by the texture of the soil horizon

Yield rating: 1 = Very unfavourable. 2 = Unfavourable. 3 = Moderately favourable. 4 = Favourable. 5 = Very favourable

Soil genesis: D = Diluvial soils of the undulating lowlands and hills; Lö = Loess soils; Al = Alluvial soils of broad river valleys including terraces and lowlands; K = Soils of the coastal regions; V = Weathered soils of solid rock and their surrounding rock masses in the mountains and hills; Vg = Rock-rich weathered soils of the high mountains

Soil quality classes: Hh = High moor; Hn = Low moor; Ls2–4 = Weak to strong sandy loam; Lt2 = Weak clay loam; Lt3 = Medium clay loam; Lts = Sandy clay loam; Lu = Silty clay loam; Sl2 = Light loamy sand; Sl3 = Medium loamy sand; Sl4 = Very loamy sand; Slu = Silty loamy sand; Ss = Pure sand; St2 = Light loamy sand; St3 = Medium loamy sand; Su2 = Light silty sand; Su3 = Medium silty sand; Su4 = Very silty sand; Tl = Clayey loam; Ts2 = Light sandy loam; Ts3 = Medium sandy loam; Ts4 = Very sandy loam; Tt = Pure loam; Tu2–4 = Weak to strong silty loam; Uls = Sandy-loamy silt; Us = Sandy silt; Ut2–4 = Weak to strong clayey silt; Uu = Pure silt

Table 2.10 Factors influencing the soil-specific yield potential (Schlutow and Schröder 2021)

Criterion	Value	Scores
Percentage of plant-available adhesive water	> 22 Vol%	5
	20–22 Vol%	4
	17–< 20 Vol%	3
	13–< 17 Vol%	2
	< 13 vol%	1
Proportion of rootable air-filled pores	> 15 Vol%	5
	12–15 vol%	4
	9–< 12 vol%	3
	5–< 9 vol%	2
	< 5 vol%	1
Pore fraction with dead water	< 5 vol%	5
	5–< 10 vol%	4
	10–< 15 vol%	3
	15–25 per cent by volume	2
	> 25 Vol%	1
Sum of dead water and air void fraction	< 20 vol%	5
	20–< 25 vol%	4
	25–< 30 vol%	3
	30–35 vol%	2
	> 35 Vol%	1
Cation exchange capacity	> 20 cmol /kg$_c$	5
	15–20 cmol /kg$_c$	4
	10–< 15 cmol /kg$_c$	3
	5–< 10 cmol /kg$_c$	2
	< 5 cmol /kg$_c$	1
Rootable depth	Deep	5
	Medium to deep	4
	Medium	3
	Medium to superficial	2
	Shoal	1
Trend towards consolidation	Soils without a tendency to podzolisation	5
	Podzoluvisol	4
	Arenosol	3
	Cambic podzol	2
	Sandy Podzol	1
Temperature-dependent humus accumulation rate	> 7 °C; ≤ 11 °C	5
	> 4.5 °C; ≤ 7 °C	4
	>1.5 °C; ≤ 4.5 °C	3
	>11 °C	2
	<1.5 °C	1

Complementary to the proportion of air voids are the proportions of water-filled pores in which the water tension due to adhesion is greater than the suction tension of the plant roots (pF > 4.2 = dead water). The proportion of very small pores with high adhesion forces is particularly high in clays (42% by volume) and zero in coarse sands.

In soils with a high proportion of medium and fine pores and a low proportion of coarse pores (silt, clay), retained water leads to a lack of air and to waterlogging due to retained water. The risk of waterlogging can therefore also be derived from the proportion of dead water pores (pF > 4.2).

Risk of dehydration:
The supply of water to plants in anhydromorphic or drained soils depends directly on the usable field capacity. While the adhesion and adsorption forces in large soil pores (e.g., in soils that consist mainly of sand) are not sufficient to form a water column in the pores, so that the precipitation water flows mainly as seepage water into the deeper soil layers and is no longer available to the plants, the very high adhesion tension against water in the narrow pores, e.g.,of silt and clay, also represents an irretrievable loss of water for the plants (permanent water loss).silt and clay, for example, also represents an irretrievable loss of water for the plants (permanent wilting point at pF > 4.2). Both soil types are therefore particularly susceptible to drying out. The combination of the dead water pore content and the air pore content results in the categorisation of the risk of drying out.

Influence of groundwater:
This criterion indicates the influence of groundwater on the plant growth of non-wetland-dependent plant species. The following applies: If the groundwater distance is less than the potential rooting depth, plant growth is restricted due to the lack of air in the soil pores. A direct groundwater influence (soil moisture) can therefore have an unfavourable influence on plant growth. A favourable influence is exerted by a groundwater table distance at which the capillary rise force specific to the soil type [closed capillary space] reaches the effective rooting depth and thus ensures sufficient soil moisture at all times. If the closed capillary space above the groundwater table generally never reaches the effective rooting depth, the non-existent influence of the groundwater is classified as "very unfavourable" in this classification. However, it also depends on the nature of the soil. For example, it can be assumed that diluvial, loessic and weathered soils are not influenced by groundwater, while alluvial and coastal soils are generally close to groundwater, i.e., the effective rooting depth of the capillary space is reached (Scheffer and Ulrich 1960).

Humus content:
The organic matter content of the mineral topsoil is largely dependent on climatic influences. The organic matter of the soil is of enormous importance, e.g., for the water storage capacity, the base sorption capacity and thus for nutrient storage and mobility. For this reason, the humus content was used as a criterion for categorising the nutrient balance (Scheffer and Ulrich 1960).

Cation exchange capacity:
The cation exchange capacity represents the potential number of exchangeable cations (calcium, magnesium, potassium, ammonium ions) and other ions (e.g.,hydrogen and aluminium ions) required for plant nutrition in the soil complex. The type and proportion of clay minerals and organic substances determine the cation exchange capacity. The cation exchange capacity of clay minerals is essentially permanent. The soil type-specific potential cation exchange capacities are highest for high clay and silt contents in the upper horizons (30 cmolc/kg for loamy, silty and pure clays) and lowest for gritty and pure sands (2 cmolc/kg) (AG Boden 2005).

Rootable depth:
Information on the rooting depth (shallow, medium or deep) can be derived indirectly from the formation and directly from the groundwater table. The soil types were therefore further subdivided into genesis types (diluvial, alluvial, weathered soils).

Tendency towards solidification:
This criterion indicates the degree of internal cohesion of horizons or layers as a result of the action of cementing substances. The higher the degree of cementation of the soil particles (e.g.,through deposits), the stronger the tendency towards consolidation. According to Hennings (1994), non-cohesive soils with a low humus content in particular tend to form cement structures with a high degree of consolidation.

To determine the soil-specific net primary productivity, the individual parameters were determined according to the soil texture (Table 2.10). If a soil profile is composed of horizons of different soil textures, the mean value of the evaluation points was calculated over the profile ($EP_{(geo-prof)}$, i.e., over all horizons of the root zone, taking into account the respective thickness of the individual horizons (depth-weighted averaging).

(c) Determination of climate-specific net primary productivity
The length of the growing season is a very important factor influencing net primary production. This is because the longer the vegetation period lasts in the year (number of days in the year with an average air temperature of ≥ 10 °C), the greater the net primary production. Good to very good growth rates are promoted by vegetation periods of 100 days (mid-montane sites) to 200 days (flat lowland sites), while in high-montane and alpine regions (60–100 days), net primary production is significantly lower than soil-specific net primary productivity. Therefore, the soil-specific net primary productivity was related to the vegetation period and classified (Table 2.10). The length of the growing season in relation to the mean annual temperature is a factor that influences net primary productivity. Precipitation also influences net primary productivity (Table 2.10).

(d) Integrative evaluation of the service potential of primary biomass production
When combining the criteria-specific evaluations (a) to (c), a ranking of the criteria was taken into account, which assumes that the individual criteria should not be included equally in a combination. The criterion of plant-physiological net primary

production (annual above-ground wood increment) was given first priority, followed by soil-water balance, nutrient balance, soil structure and climate influence.

The assessment of the plant physiological net primary production was then carried out according to the following formula:

$$NPP = P + ((W - P) * 0.5) + ((N - P) * 0.25) + ((G - P) * 0.125) + ((K - P) * 0.06)$$

With:

NPP = evaluation score for the total primary productivity of the biomass
P = evaluation score for plant physiological net primary production (annual above-ground wood growth)
W = Evaluation grade soil water balance
N = evaluation score for the nutrient balance
G = Evaluation score for the soil structure
K = Evaluation score for the climate

The weighting according to this formula is based on the "Soil Quality Rating" (SQR) (Müller et al. 2007). However, this only applies to Germany and possibly to Central Europe. In other European regions where climatic factors limit primary production to a greater extent, the weighting must be adjusted accordingly.

2.4 Carbon Sequestration Service

Carbon (C) is stored in the living biomass and in the soil organic matter. However, the proportion of carbon in the living biomass is around one third of that in the soil (Schlesinger 1990). In the following, the carbon storage capacity in the soil is therefore considered in simplified terms. Part of the C stock from dead biomass is located in the (still) undecomposed leaf/needle litter in the humus layer (on average approx. 18% in German forest soils according to Grüneberg et al. 2017). The largest C reserve (59%, Grüneberg et al. 2017) is located in the mineral soil layer up to 30 cm. Here, the humic substances are incorporated into long-term stable organic-mineral complexes (e.g.,clay-humus complexes).

The following criteria are of particular importance for categorising the service of ecosystems for carbon sequestration (Grüneberg et al. 2017):

(a) Clay content
(b) Mass of litter and crop residues
(c) Decomposability of litter and crop residues
(d) Annual mean temperature and annual precipitation total
(e) Water content of the soil
(f) pH value.

The ecosystem service was assessed by assigning levels (0–5: very low to very good) to a classified expression of the characteristics listed above under a) to f).

(a) Clay content

Since the C stock in the organic layer is on average only about one fifth of the C stock in the mineral soil, the clay content is significantly positively correlated with the C stock of the soil profiles (Grüneberg et al. 2017). However, the clay content is also a criterion that influences soil fertility and thus the amount of needle and leaf litter, the soil water content, the effective cation exchange capacity and the pH value and thus the activity of humus decomposers, among other things. The clay content is the strongest factor influencing carbon storage (Grüneberg et al. 2017; Grabe et al. 2003; Wiesmeier et al. 2013). The categorisation according to AG Boden (2005, p. 142) was applied (Table 2.11).

(b) Mass of litter and crop residues

Carbon is predominantly stored in the soil in organic form. It comes partly from non-mineralised components, in particular from undecomposed or only partially decomposed components of plant litter and crop residues (leaves, needles, bark, branches, coarse and fine roots). In forests, litter mass depends on soil fertility, but also on the main tree species, with some stand-forming tree species favouring soils with a certain nutrient content and thus serving as indicators of soil fertility. The evaluation of litter mass production was based on Jacobsen et al. (2002), Madl (2017), Schachtschabel et al. (1998) and Scheffer and Ulrich (1960) (Table 2.12 in conjunction with Tables 2.8, 2.9 and 2.10).

(c) Decomposability of litter and crop residues

The decomposability of plant residues also depends on the chemical composition of the litter. A low C/N ratio and a high pH value in the fresh litter promote rapid decomposition. A well-studied criterion for degradability is the degradation time of the litter and crop residues according to Blum et al. (2018; Table 2.13).

Table 2.11 Ordinally categorising the clay content

Parameters	Scores
Clay content >60%	5
Clay content >30–60%	4
Clay content >20–30%	3
Clay content >10–20%	2
Clay content >5–10%	1
Clay content ≤5%	0

Table 2.12 Ordinally categorising litter mass productivity

Parameters	Scores
> 5.5 tonnes dry matter ha^{-1} a^{-1}	5
> 4.5 to ≤5.5 tonnes dry matter ha^{-1} a^{-1}	4
> 3.5 to ≤4.5 tonnes dry matter ha^{-1} a^{-1}	3
> 2.5 to ≤3.5 tonnes dry matter ha^{-1} a^{-1}	2
≤ 2.5 tonnes of dry matter ha^{-1} a^{-1}	1

Table 2.13 Ordinal classification of the decomposition time

Parameters	Scores
≥ 3 to–> 4 years	5
≥ 2 to–≤ 4 years	4
≥ 1 to–≤ 3 years	3
≥ 0.5 to–≤ 2 years	2
≤ 0.5 to–≤ 1 year	1

Table 2.14 Criteria, parameters and scores for ordinally categorising carbon sequestration service

Parameters	Scores
≤ 0.05 to 0.15 (m^3 water m^{-3} soil)	1
0.14 to 0.3 (m^3 water m^{-3} soil)	5
0.3 to 0.42 (m^3 water m^{-3} soil)	4
0.42 to 0.66 (m^3 water m^{-3} soil)	3
0.5 to 0.97 (m^3 water m^{-3} soil)	2
> 0.97 (m^3 water m^{-3} soil)	0

(d) Water content of the soil

A good water supply in combination with a good oxygen supply promotes the activity of the decomposers and thus the formation of stable clay-humus complexes in the mineral soil. Drought, on the other hand, leads to a reduction in activity. Anaerobic conditions in the soil also inhibit the mineralisation of the litter. Although decomposition also takes place in the absence of oxygen, as methanogenic microorganisms, for example, take over the mineralisation, the decomposition process under anaerobic conditions is considerably slower and usually incomplete (Schachtschabel et al. 1998). The inhibition of decomposition in water-saturated soils can assume such proportions that peat layers form from less decomposed litter, which in turn accumulate a high stock of organically bound carbon (Succow and Joosten 2001). The influence of the water content on the C stock was classified according to AG Boden (2005:350) (Table 2.14).

(e) Annual mean temperature and annual precipitation total

While the carbon stock in the humus increases at high water contents and lower average annual temperatures, the carbon accumulation rate in the mineral soil, where the largest proportion of the C stock is located, decreases at high water contents and low temperatures. The rate of mineralisation of organic matter and the subsequent formation of stable organic-mineral complexes is controlled by temperature and precipitation, among other factors. The biological activity of the mineralising soil organisms increases with rising soil temperature. The soil temperature in the biologically active topsoil is rarely measured and depends on the average annual air temperature (Table 2.15). The activity of humus-degrading soil organisms is inhibited by anaerobic conditions in the soil. High precipitation leads to anaerobic conditions more frequently and over longer periods of time (Blum et al. 2018).

Table 2.15 Criteria, parameters and scores for the assessment of the annual mean temperature/annual precipitation sum

Parameters	Scores
−4 to 8 °C/1065 to 2710 mm	1
4.7 to 11.1/745 to 1291 mm	2
5.5 to 11.1/514 to 854 mm	3
8.1 to 12/380 to 632 mm	4
> 12/< 400 mm	5

Table 2.16 Criteria, parameters and scores for ordinally categorising the $pH_{(H2O)}$ values

Parameters	Scores
< 3.8	1
4.2–3.8	2
5.0–4.2	3
6.2–5.0	4
8.6–6.2	5

(f) pH value

In different pH ranges, the organism community in the soil is made up of different species or species groups that develop different decomposition intensities. If the pH value measured in the water ($pH_{(H2O)}$) is <4.2, for example, earthworms and bristle worms are no longer viable. In addition to decomposing humus, they are also responsible for combining humic and mineral substances and transporting them from the humus-rich topsoil to the upper mineral soil layer. On the other hand, the pH value is usually strongly correlated with the base saturation, so that a high pH value is also an indicator of a good nutrient cation supply for the humus decomposers. The lower the pH value, the lower the activity of the decomposers (Leuschner et al. 2013). However, high destruent activity is a prerequisite for the formation of organo-mineral complexes in the mineral soil, in which the highest C content is stored (Table 2.16).

(g) Integrative assessment of the carbon sequestration capacity of ecosystems

When integrating the criteria-specific classification rankings (a) to (f), a ranking of the criteria was taken into account, which assumes that the individual criteria should not be weighted equally. The clay content criterion was given first priority, followed by the production of litter mass. The production of litter mass results only from the maximum possible litter mass production in terms of plant physiology, relativised on the basis of the yield potential of the soil. In third place are the factors influencing the decomposition activity of the humus decomposers. The influencing factors pH value, climate, volumetric water content in the soil and the decomposition time are weighted equally and summarised to an arithmetic mean value.

In order to classify the clay content T, additions or deductions were made according to the following equation:

$$K = T + \left(\left(\left(\left(EP_{geo} - prof - 1 \right) / \left(5 - 1 \right) \right) * S_{pot} \right) - T \right) * 0,5 \right)$$
$$+ \left(\left(\left(pH + J + W + Z \right) / 4 - T \right) * 0,25 \right)$$

With:

K = Overall assessment of carbon sequestration service
T = Evaluation of the total tonal content
Spot = mass productivity of the litter
EP_{geo} -prof = yield potential of the soil profile
$pH_{(H2O)}$ = pH value (pH) measured in water
J = Climate (annual precipitation sum/annual mean temperature)
W = water content of the soil
Z = decomposition time

The weighting of the criteria for the assessment of carbon storage capacity was based on the analysis and evaluation of the Soil Condition Survey in Germany 2006–2008 (Grüneberg et al. 2017).

2.5   Rough Estimates of Ordinal Classification of Further Ecosystem Services

A rule-based assessment of other ecosystem services included in CICES according to Haines-Young and Potschin (2018) was also carried out using expert judgement. However, this was done at a rough level and is only intended as a suggestion for further consolidation (Table 2.17). In addition, a classification of all ecosystem services relevant in Germany that is as complete as possible is necessary as a prerequisite for summarising ecosystem types into ecosystem type classes (Sect. 3.1.1). However, a rough estimate is sufficient for this purpose.

Table 2.17 Rules for the approximate assessment of other ecosystem services

Ecosystem services	Criterion	Parameters	Scores
Erosion control function—stabilisation of solids (soil, sand, snow, etc.)	Vegetation cover	Deciduous forests (cover >80%)	5
		Mixed deciduous/coniferous forests (cover >80%)	5
		Coniferous forests (cover 70 to 80%)	4
		Deciduous forest close to the natural forest on soils with high and rich nutritional value (cover 60- < 70%).	3
		Mixed coniferous and deciduous forests close to the natural forest on nutrient-rich soils (cover 60- < 70%).	3
		Coniferous wood near the natural forest on soils with strong and rich nutrients (cover 50- < 60%).	2
		Mixed conifer/deciduous trees close to the natural forest on soils with low to medium nutritional value (cover 50 to 60%).	2
		Near-natural coniferous stand on soils with low to medium nutritional value (cover <50%).	1
Flood protection function	Storage capacity of the floor	Soils with thick peat layers (humus content >30%)	5
		Soils with high clay/silt content (> 50%) and high humus content (> 15%)	4
		Soils with high clay/silt content (> 50%) and low humus content (< 15%).	3
		Soils with low clay/silt content (< 50%) and high humus content (> 15%).	2
		Soils with low clay/silt content (< 50%) and low humus content (< 15%).	1
Groundwater recharge/ drinking water Water supply function.	Annual precipitation	Temperate boreal climate	5
		Temperate sub-oceanic climate, partly mountain climate	4
		Temperate suboceanic to temperate subcontinental climate, partly mountain climate	3
		Temperate Central European climate	2
		Temperate subcontinental climate	1

(continued)

Table 2.17 (continued)

Ecosystem services	Criterion	Parameters	Scores
Groundwater protection function/ safeguarding drinking water quality	Permeability of the soil	Fine and medium gravel (permeability coefficient: $K_f = 7 \times 10^{-3}$ m/s)	5
		Medium sands ($K_f = 1 \times 10^{-4}$ m/s)	4
		Silty subsoils and interbedded sands, sandy deep loamy soils (Kf = strongly fluctuating from 5×10^{-5} to 1×10^{-6} m/s).	3
		Niederungstorf (Riedfen) ($K_f = 6 \times 10^{-5}$ m/s)	3
		Degraded lowland peat (mulm) in drained lowlands and siltation zones ($K_f = 3 \times 10^{-6}$ m/s)	2
		Silt with sand deposits (Kf = strongly fluctuating from 1×10^{-6} to 1×10^{-8} m/s)	1
		Silt and clay soils ($K_f < 1 \times 10^{-8}$ m/s)	0
Nature experience and recreational function	Diversity/ renewal— capacity	High diversity, high naturalness, high regeneration capacity (near-natural deciduous and mixed forests on river banks or in floodplains)	5
		High diversity, high naturalness, high regeneration capacity (near-natural deciduous and mixed forests)	4
		Medium diversity, low naturalness, high regeneration capacity (deciduous or mixed forests with few tree species)	3
		Low diversity, low closeness to nature, high regeneration capacity (coniferous forests)	2
		High diversity, high naturalness, very low regeneration capacity and forests that are difficult to access (e.g., near-natural, moist)	1
		Very low regenerative capacity and inaccessible (e.g., moorland forests).	0
Aesthetic perception, nature education, natural heritage	Naturalness/ individuality	Very high naturalness, special character (moor and wet forests, rocky dry forests)	5
		High degree of naturalness, special character (e.g., near-natural deciduous and mixed forests)	4
		High naturalness, medium specificity (e.g., near-natural coniferous forests in the mountains)	3
		Medium naturalness, no characteristic (deciduous and mixed forests)	2
		Low naturalness, no characteristic (deciduous or mixed forests with few tree species)	1
		Very low naturalness, no characteristic (coniferous forests)	0

(continued)

Table 2.17 (continued)

Ecosystem services	Criterion	Parameters	Scores
Preservation function of cultural heritage, legacy for future generations	Rarity/cultural-historical value	Currently rare and threatened by changes in land use, evidence of former extensive land use (e.g., hornbeam-rich woodland)	5
		Currently rare semi-natural forests that are threatened by changes in management (e.g., semi-natural wetlands and moorland forests).	4
		Currently rare forests and forests threatened by changes in management (e.g., wetlands and bog forests)	3
		Currently rare but not endangered semi-natural forests	2
		Currently not rare and not threatened deciduous and mixed forests, deciduous forests	1
		Currently not rare and not endangered coniferous forests	0

References

AG Boden – Arbeitsgruppe Boden (2005) Bodenkundliche Kartieranleitung [Soil mapping guide]. 5. Aufl., Bundesanstalt für Geowissenschaften und Rohstoffe und den Geologischen Landesämtern der Bundesrepublik Deutschland (Ed.), Hannover

Amberger A. (1988) Pflanzenernährung—Ökologische und physiologische Grundlagen, Dynamik und Stoffwechsel der Nährelemente [Plant nutrition—ecological and physiological basics, dynamics and metabolism of nutrient elements]. Stuttgart: 3. überarb. Aufl., Verlag Eugen Ulmer, S. 118 ff

Balla S, Uhl R, Schlutow A, Lorentz H, Förster M, Becker C, Scheuschner Th, Kiebel A, Herzog W, Düring I, Lüttmann J, Müller-Pfannenstiel K (2013) Untersuchung und Bewertung von straßenverkehrsbedingten Nährstoffeinträgen in empfindliche Biotope. Hrsg.: BMVBS—bundesministerium für Verkehr, Bauwesen und Städtebau. Endbericht zum FE-Vorhaben 84.0102/2009 im Auftrag der Bundesanstalt für Straßenwesen [Investigation and assessment of road traffic-induced nutrient inputs into sensitive biotopes. Final report on FE project 84.0102/2009 on behalf of the Federal Highway Research Institute], written by = Forschung Straßenbau und Straßenverkehrstechnik, Heft 1099, BMVBS Abteilung Straßenbau, Bonn. 362 pp.

BArtSchV (2005) Verordnung zum Schutz wild lebender Tier- und Pflanzenarten Anlage 1 (Bundesartenschutzverordnung). [Federal Species Protection Ordinance (BArtSchV 2005) Annex 1). https://www.gesetze-im-internet.de/bartschv_2005/anlage_1.html

Blum W, Schad P, Nortcliff S (2018) Essentials of soil science. Soil formation, functions, use and classification (world Reference Base, WRB). Borntraeger Science Publishers, Stuttgart. 171 pp

BMU (Bundesministerium für Umwelt, Naturschutz und Reaktorsicherheit) (2007) Nationale Strategie zur Biologischen Vielfalt (vom Bundeskabinett am 07.11.2007 beschlossen), Oktober 2007 [BMU (Federal Ministry for the Environment, Nature Conservation and Nuclear Safety Germany) (2007): National Strategy on biological Diversity. Bonifatius GmbH Paderborn, 180 p. https://www.bfn.de/fileadmin/ABS/documents/Biodiversitaetsstragie_englisch.pdf

Böckmann T (1990) Wachstum und Ertrag der Winterlinde [growth and yield of the littleleaf linden] (Tilia cordata mill) in Nordwestdeutschland. Dissertation University, Göttingen

European Commission (1992) Council Directive 92/43/EEC of 21 May 1992 on the conservation of natural habitats and of wild fauna and flora. Eur-Lex. Retrieved 9 March 2020

European Parliament (2009) Directive 2009/147/EC of the European Parliament and of the Council of 30 November 2009 on the conservation of wild birds Council Directive 92/43/EEC of 21 May 1992 on the conservation of natural habitats and of wild fauna and flora (FFH Annexes II and IV). https://eur-lex.europa.eu/legal-content/EN/TXT/?uri=CELEX%3A32009L0147

Fröhlich, Sporbeck (2002) Leitfaden zur Erstellung und Prüfung Landschaftspflegerischer Begleitpläne zu Straßenbauvorhaben in Mecklenburg-Vorpommern. Erläuterungsbericht. Erstellt im Auftrag des Landesamtes für Straßenbau und Verkehr Mecklenburg-Vorpommern [Guidelines for the preparation and examination of landscape conservation plans for road construction projects in Mecklenburg-Western Pomerania. Explanatory report. Created on behalf of the State Office for Road Construction and Transport Mecklenburg-Vorpommern]. Bochum, Schwerin, 164 S

Grabe M, Kleber M, Hartmann K-J, Jahn R (2003) Preparing a soil carbon inventory of Saxony-Anhalt, Central Germany using GIS and the state soil data base SABO_P. J Plant Nutr Soil Sci 166:642–648

Grüneberg E, Riek W, Schöning I, Evers J, Hartmann P, Ziche D (2017) Kohlenstoffvorräte und deren zeitliche Veränderungen in Waldböden [Carbon stocks and their temporal changes in forest soils]. In: Wellbrock, N., Bolte, A., Flessa, H. (Eds.): Dynamics and spatial patterns of forest locations in Germany. Results of the soil condition survey in the forest 2006–2008 [Dynamics and spatial patterns of forest locations in Germany. Results of the soil condition survey in the forest 2006–2008]. Thünen Report 43, I-183 to I-209

Haines-Young R, Potschin MB (2018) Common international classification of ecosystem services (CICES) V5.1and guidance on the application of the revised. Structure. Available from www. cices.eu

Hennings V (1994) Methodendokumentation Bodenkunde—auswertungsmethoden zur Beurteilung der Empfindlichkeit und Belastbarkeit von Böden [Method documentation soil science—evaluation methods for assessing the sensitivity and resilience of soils]. Geologisches Jahrbuch, Reihe F, Schweizerbarthsche Verlagsbuchhandlung Stuttgart 13:194

Jacobsen C, Rademacher P, Meesenburg H, Meiwes KJ (2002) Element-Gehalte in Baum-Kompartimenten: Literatur-Studie und Datensammlung [element contents in tree compartments: literature study and data collection]. In: Niedersächsische Forstliche Versuchsanstalt, Report; self-publishing by Forschungszentrum Waldökosysteme der Universität Göttingen, Germany, pp 1–80

Jüttner (1955) Ertragstafeln der Stiel- und Traubeneiche [Yield tables of pedunculate and sessile oak]. In: Schober R (ed) Ertragstafeln wichtiger Baumarten bei verschiedenen Durchforstungen [Yield tables of important tree species in different thinnings]. Verlag Sauerländer, Frankfurt a. M

Leuschner C, Wulf M, Bäuchler P, Hertel D (2013) Soil C and nutrient stores under scots pine afforestation compared to ancient beech forests in the German Pleistocene: the role of tree species and forest history. For Ecol Manag 310:405–415

Madl P (2017) Vorlesungsskript Bodenökologie, gelesen von W. Strobl, Universität Salzburg [Lecture notes on soil ecology, read by W. Strobl, University of Salzburg]. http://biophysics. sbg.ac.at/transcript/boden.pdf

Müller L, Schindler U, Behrendt A, Eulenstein F, Dannowski R (2007) The Muencheberg Soil Quality Rating (SQR). http://www.zalf.de/de/forschung/institute/lwh/mitarbeiter/lmueller/ Documents/field_mueller.pdf. Accessed 25 July 2019

Schachtschabel P, Auerswald K, Brümmer G, Hartke KH, Schwertmann U (1998) Lehrbuch der Bodenkunde [Textbook of soil science]. Verlag Ferdinand Enke, Stuttgart

Scheffer F., Ulrich B. (1960): Humus und Humusdüngung [Humus and humus fertilization]. Zweite, völlig neu bearbeitete Auflage [second, completely reworked edition]. Band I: Morphologie, Biologie, Chemie und Dynamik des humus. Mit 45 Abbildungen und 39 Tabellen. 1960. VII, 266 seiten [volume I: Morphology, biology, chemistry and dynamics of humus. With 45 figures and 39 tables. 1960. VII, 266 p.]

Schlesinger WH (1990) Evidence from chronosequence studies for a low carbon-storage potential of soils. Nature 348:232–234

Schlutow A, Scheuschner T (2023) Determination of critical loads for Eutrophying and acidifying air pollutant inputs for the protection of near-natural ecosystems in Germany. Atmos 14(2):383. https://doi.org/10.3390/atmos14020383

Schlutow A, Schröder W (2021) Rule-based classification and mapping of ecosystem services with data on the integrity of forest ecosystems. Environ Sci Eur 33:50. https://doi.org/10.1186/s12302-021-00481-3

Schlutow A, Bouwer Y, Nagel H-D (2018) Bereitstellung der Critical Load Daten für den Call for Data 2015–2017 des Coordination Centre for Effects im Rahmen der Berichtspflichten Deutschlands für die Konvention über weitreichende grenzüberschreitende Luftverunreinigungen (CLRTAP). Im Auftrag des UBA, Abschlussbericht Projekt-Nr. UBA/43848. [Provision of critical load data for the Call for Data 2015–2017 of the Coordination Centre for Effects in the context of Germany's reporting obligations for the Convention on Long-Range Transboundary Air Pollution (CLRTAP). On behalf of UBA, Final Report Project no. UBA/43848]. https://www.umweltbundesamt.de/publikationen/critical-load-daten-fuer-die-berichterstattung-2015

Schlutow A, Schröder W, Scheuschner T (2021) Assessing the relevance of atmospheric heavy metal deposition with regard to ecosystem integrity and human health in Germany. Environmental sciences. Europe 33(1). https://doi.org/10.1186/s12302-020-00391-w

Schlutow A, Kraft P, Scheuschner T, Schlutow M, Schröder W (2024) Bioindication for ecosystem regeneration towards natural conditions—the BERN data base and BERN model. Environmental sciences. Europe 36(1). https://doi.org/10.1186/s12302-023-00826-0. BERN database (open source) under. https://github.com/bern-model/BERN

Schober R (1975) Ertragstafeln wichtiger Baumarten bei verschiedenen Durchforstungen [Yield tables of important tree species in different thinnings]. J. D. Sauerländer's Verlag, Frankfurt/M

Schober R (1987) Ertragstafeln wichtiger Baumarten [Yield tables of important tree species]. J. D. Sauerländer's Verlag, Frankfurt/M

Schwappach H (1912) Ertrags-Schätztafeln für Forstbestände [Yield estimation tables for forest stands]. Archiv der Forstwissenschaft Eberswalde, unveröffentlicht

Succow M, Joosten H (2001) Landschaftsökologische Moorkunde [Landscape ecology of peatlands], 2nd edn. Schweizerbart'sche Publishersbuchhandlung. Stuttgart. 622 pp

Wiedemann F (1936) Ertragstafeln der Fichte [Yield tables of spruce]. In: Schober (ed) Ertragstafeln wichtiger Baumarten bei verschiedenen Durchforstungen [Yield tables of important tree species in different thinnings]. Verlag Sauerländer, Frankfurt a. M

Wiedemann F (1943) Ertragstafeln der Kiefer [Yield tables of pine]. In: Schober (ed) Ertragstafeln wichtiger Baumarten bei verschiedenen Durchforstungen [Yield tables of important tree species in different thinnings]. Verlag Sauerländer, Frankfurt a. M

Wiesmeier M, Prietzel J, Barthold F, Spörlein P, Geuß U, Hangen E, Reischl A, Schilling B, von Lützow M, Kögel-Knabner I (2013) Storage and drivers of organic carbon in forest soils of Southwest Germany (Bavaria)—implications for carbon sequestration. For Ecol Manag 295:162–172

Wimmenauer K (1919) Wachstum und Ertrag der Esche [Ash tree growth and yield]. Allgemeine Forst- und Jagd-Zeitschrift 90:9–17. & 37-40

Chapter 3
Data Basis and Mapping Results of Ecosystem Services at Different Levels

Abstract In large regions with a large number of different ecosystem types, the application of the rules for the assessment of ecosystem services first requires a practical grouping of the types into ecosystem type classes. This chapter presents a method for grouping forest ecosystem types in Germany, which can also be applied to Central Europe.

The data basis for the Germany-wide assessment of the habitat function was derived from the BERN model (Bioindication for Ecosystem Regeneration towards Natural conditions) (Schlutow A, Kraft P, Scheuschner T, Schlutow M, Schröder W, Environ Sci Eur 36(1), 2024). The BERN database documents reference data on typical site parameters for the occurrence of plant communities.

The assessment of the biomass primary productivity and carbon sequestration service of ecosystem type classes was based on the spatial distribution of tree species in current forests and woodlands in Germany (BERN database in connection with Corine Land Cover) as well as on data on soil properties (soil overview map of Germany 1:1 million—BÜK1000N) and climate (30-year average 1991–2020).

For the assessment of the Kellerwald National Park, 105 vegetation surveys were the basis for the indication of soil parameters.

The derivation of climate change-adapted lead forest communities in Saxony is based on forest site mapping combined with climate projections up to 2070 and the BERN database.

The effects of acidifying and eutrophying air pollution impacts and climate change on biodiversity were already assessed by example of a Long-Term Ecological Research site (LTER) using timeseries of atmospheric sulphur and nitrogen deposition from year 1880 to year 2070 as well as timeseries of temperature and precipitation.

Keywords Ecosystem-type classes · Ecosystem services maps · Acidification effects · Eutrophication effects · Climate effects · BERN database

© The Author(s), under exclusive license to Springer Nature
Switzerland AG 2024
A. Schlutow, W. Schröder, *Climate Change and Atmospheric Deposition as Drivers of Forest Ecosystem Integrity and Services*, SpringerBriefs in Environmental Science, https://doi.org/10.1007/978-3-031-67103-6_3

3.1   National Mapping

3.1.1   Rule-Based Classification of Forest Ecosystem Type Classes

In order to be able to carry out a regionalisation of the assessment of ecosystem services, the regionalisation of ecosystem type classes is first required as a basis. For the ecosystem type classes in Germany (representative of Central Europe), the following criteria were used to define the ecosystem type classes.

Ecosystem type classes should be classified with the aim of grouping together ecosystem types with similar vegetation types (species composition, structure and utilisation) and similar site parameters if they also have qualitatively and quantitatively comparable ecosystem service potentials. For reasons of clarity, as few ecosystem type classes as possible, but as many as necessary, should be differentiated in order to ensure a clear distinction in terms of site characteristics, vegetation type and ecosystem integrity. Forest ecosystem type classes can be clearly assigned to site types in Germany if they are defined by a combination of ecoclimatic zones, soil moisture status and nutrient cycling type (Schlutow and Schröder 2021). This is described below.

The climate classification was based on plant geographic distribution patterns of near-natural forest plant communities or their main tree species and the allocation of ranges of mean annual temperature and total annual precipitation (Table 3.1). In

Table 3.1 Ecoclimatic zones based on mean annual temperature and total annual precipitation (according to Balla et al. 2013) using the distribution of the most important tree species in Germany

Ecoclimatic zones		Annual average temperature [°C]	Annual precipita-tion sum [mm a^{-1}]	Climatic water balance [mm/ Veg. month]	Length of vegetation period [d > 10 °C a^{-1}]
1	Temperate boreal climate (mountain pine, larch, spruce, fir)	−4–8	1065–2710	>36	90–140
2	Temperate sub-oceanic climate, partly Borel climate (beech, partly fir)	4.7–11.1	745–1291	2–69	140–190
2–3 (2.5)	Temperate suboceanic to temperate subcontinental climate, partly boreal climate (beech)	5.5–11.1	514–854	−23–2	140–190
3	Temperate Central European to subcontinental climate (oaks, pines, hornbeams, lime trees).	8.1–12	380–632	−47–−23	190–220

this way, the climate classification can be retraced at any time using the original DWD data (DWD 2012) and updated if necessary (e.g., 1991–2020, DWD 2021).

The soil moisture was also categorised according to plant physiological aspects. The volumetric water content in the topsoil [m^3 water/m^3 soil] refers to the field capacity range in the effective rooting zone. The lower range limit is given in Table 3.2 The lower range limit for the anhydromorphic soil forms results from the water content at the permanent wilting point at pF = 4.2 (AG Boden 2005, p. 350), the upper range limit at saturated field capacity, i.e., at pF = 1.8. The range in Table 3.2 for the hydromorphic soil forms results from the water content of pF 0.5 to 1.8.

The classification of the nutrient cycle types of German forest ecosystems was based on the C/N ratio in the topsoil (averaged over humus topsoil +5 cm mineral topsoil) and the base saturation (averaged over the entire root zone) (Schulze 1998; Schlutow and Schröder 2021; Table 3.3).

The grouping of forest ecosystem types by main tree species is based on the mapping of forest types in the Corine Land Cover database for Germany (UBA 2015;

Table 3.2 Moisture content based on the volumetric water content according to Schulze (1998)

Moisture classes		Volumetric water content, annual average [m^3 m^{-3}]
1	Dry	≤ 0.05–0.15
2	Moderately dry to fresh	0.14–0.3
4	Moist, moist	0.3–0.42
5	Wet	0.42–0.66
6	Alternating dry to fresh, flooded	0.21–0.3
7	Damp flooding	0.3–0.52
8	Very wet	0.5–0.97

Table 3.3 Classes of nutrient cycle types based on C/N ratio and base saturation according to Schulze (1998) (adapted by Schlutow and Schröder 2021)

Classes of nutrient cycle types		C/N [%/%]	Base saturation [%]
1	Bad	29–50	3–15
2	Pretty bad	26–33	10–20
3	Moderately nutritious	17–26	15–30
4	Moderately nutritious, carbonated	17–21	> 90
5	Strong and nutritious	13–18	30–50
6	Strong, nutritious, carbonate-containing	13–15	> 90
7	Rich in nutrients	11–14	50–80
8	Nutrient-rich, carbonate-rich	8–11	80–100
9	Poor, organic	30–60	< 26
10	Pretty poor, organic	26–36	< 26–52
11	Moderately nutritious, organic	20–26	< 26–52
12	Vigorous, nutritious, organic	13–20	> 26
13	Nutrient-rich, carbonate-containing, organic	< 10–30	> 90
14	Fairly poor, carbonate-containing, organic	20–31	> 52

Table 3.4). The comparison of the forest types from Corine Land Cover with the distribution of semi-natural vegetation from the BERN model (Schlutow et al. 2024) shows where natural coniferous forests occur and where coniferous forests have been afforested to replace natural deciduous forests. Accordingly, a distinction was made in the classification and mapping of ecosystem type classes between natural or semi-natural "forests" and non-natural "forests" (Table 3.4).

The regional distribution of the 78 ecosystem type classes is shown in Fig. 3.1.

The regionalisation of the 78 ecosystem type classes in Germany (Table 3.4) was carried out by intersecting the forest type mapping (Corine Land Cover 2012, UBA 2015) with the 1:1 million soil overview map (BGR 2014) and with the climate data of the German Weather Service (DWD 2012). The soil moisture, the C/N ratio and the base saturation can be assigned to the polygons of the soil forms from the database of the 1:1 million soil overview map (BGR 2014). The DWD raster maps contain 1x1km^2 grids with the annual mean temperature and the annual precipitation sum in the long-term average 1981–2010. The Corine Land Cover map contains the forest type information: coniferous, deciduous or mixed forest.

3.1.2   Evaluation of the Ecosystem Services of the Forest Ecosystem Type Classes

The following objectives were pursued in this section:

1. The methodology developed to determine ecosystem service potentials should be applicable for the whole of Germany and, if possible, for the whole of Central Europe.
2. The user should be given the opportunity to assess the service in specific individual cases in a comprehensible and reproducible manner using measurable parameters. The following method serves the user in particular to link status information in order to estimate the deviation of the current or expected future value from the original status, e.g., due to climate change and atmospheric nitrogen deposition. In this way, the user can identify the influencing factor that significantly determines the extent of the deviation of the current service score from the score of unaffected ecosystems and build the strategy for effective avoidance or restoration measures on this knowledge.

The application of the criteria according to Sects. 2.2, 2.3 and 2.4 to the ecosystem type classes is shown in Table 3.11. The following maps and data sources form the basis for the mapping of ecosystem services in Germany:

Table 3.4 78 Wood/forest ecosystem type classes based on abiotic site factors and forest types

Ecoclimatic zones	Condition of the soil moisture	Type of nutrient cycle	Forest type (CORINE Land Cover)	ID	Class of the ecosystem type forest/wood
3	1	1	Conifer	1	Central European to subcontinental, dry, nutrient-poor pine forest
2	1	1	Mixed	2	Suboceanic, dry, nutrient-poor pine forest
3	1	1	Mixed	3	Central European to subcontinental, dry, nutrient-poor pine wood
2	2	1	Conifer	4	Suboceanic, moderately dry to fresh, nutrient-poor pine forest
2	2	1	Conifer	5	Suboceanic, moderately dry to fresh, poor larch forest
2	2	1	Deciduous	6	Suboceanic, moderately dry to fresh, nutrient-poor beech forest / woodland
2	2	1	Mixed	7	Suboceanic, moderately dry to fresh, nutrient-poor spruce wood
2	2	1	Conifer	8	Suboceanic, moderately dry to fresh, nutrient-poor spruce forest
2	4	1	Conifer	9	Suboceanic, moist, nutrient-poor pine forest
2	4	2	Conifer	10	Suboceanic, moist, fairly nutrient-poor pine forest
3	4	1	Mixed	11	Central European to subcontinental, moist, nutrient-poor spruce wood
3	1	2	Deciduous	12	Central European to subcontinental, dry, fairly nutrient-poor sessile oak wood
3	1	2	Conifer	13	Central European to subcontinental, dry, fairly nutrient-poor pine forest
2	2	2	Deciduous	14	Suboceanic, moderately dry to fresh, fairly nutrient-poor beech forest/woodland
1	2	2	Deciduous	15	Moderate boreal climate, moderately dry to fresh, quite nutrient-poor beech wood/forest
2	4	2	Deciduous	16	Suboceanic, moist, fairly nutrient-poor English oak forest
2	5	2	Mixed	17	Suboceanic, moist, fairly nutrient-poor fir forest
3	1	3	Conifer	18	Central European to subcontinental, dry, moderately nutrient-rich pine forest
3	1	3	Deciduous	19	Central European to subcontinental, dry, moderately nutritious sessile oak wood

(continued)

Table 3.4 (continued)

Ecocli-matic zones	Condition of the soil moisture	Type of nutrient cycle	Forest type (CORINE Land Cover)	ID	Class of the ecosystem type forest/wood
2	2	3	Conifer	20	Suboceanic, moderately dry to fresh, moderately nutrient-rich pine forest
3	2	3	Conifer	21	Central European to subcontinental, moderately dry to fresh, moderately nutrient-rich pine forests
2	4	3	Deciduous	22	Suboceanic, moist, moderately nutrient-rich oak forest
2	1	4	Conifer	23	Suboceanic, dry, moderately nutrient-rich, carbonate-rich pine forests of the Alpine valleys
2	2	3	Deciduous	24	Suboceanic, moderately dry to fresh, moderately nutrient-rich beech forest/wood
3	2	3	Deciduous	25	Central European to subcontinental, moderately dry to fresh, moderately nutritious sessile oak-beech wood
3	2	3	Deciduous	26	Central European to subcontinental, moderately dry to fresh, moderately nutrient-rich lime-hornbeam wood
1	2	3	Deciduous	27	Moderate boreal climate, moderately dry to fresh, moderately nutrient-rich beech wood/forest stand
2	2	3	Conifer	28	Suboceanic, moderately dry to fresh, moderately nutrient-rich spruce forest
2	2	3	Mixed	29	Suboceanic, moderately dry to fresh, moderately nutritious spruce-fir wood
2	2	3	Conifer	30	Suboceanic, moderately dry to fresh, moderately nutrient-rich Douglas fir forest
2	2	3	Mixed	31	Suboceanic, moderately dry to fresh, moderately nutrient-rich fir-beech forest
2	2	3	Deciduous	32	Suboceanic, moderately dry to fresh, moderately nutritious English oak-beech wood
2	2	3	Deciduous	33	Suboceanic, moderately dry to fresh, moderately nutrient-rich oak forest
1	4	3	Mixed	34	Temperate boreal climate, moist, moderately nutrient-rich fir-beech forest

(continued)

Table 3.4 (continued)

Ecocli-matic zones	Condition of the soil moisture	Type of nutrient cycle	Forest type (CORINE Land Cover)	ID	Class of the ecosystem type forest/ wood
2	4	3	Mixed	35	Suboceanic, moist, moderately nutrient-rich fir-beech forest
2	4	3	Deciduous	36	Suboceanic, moist, moderately nutrient-rich oak-hornbeam forest
2	5	3	Mixed	37	Suboceanic, moist, moderately nutritious fir wood
2	1	4	Conifer	38	Suboceanic, dry, moderately nutrient-rich, carbonate pine forest
2	2	4	Conifer	39	Suboceanic, moderately dry to fresh, moderately nutrient-rich carbonate spruce forest
3	1	5	Deciduous	40	Central European to subcontinental, dry, fast-growing, nutritious sycamore oak wood
3	2	5	Deciduous	41	Central European to subcontinental, moderately dry to fresh, vigorous, nutrient-rich hornbeam forest
3	2	5	Deciduous	42	Central European to subcontinental, moderately dry to fresh, vigorous, nutrient-rich winter lime wood
1	2	5	Mixed	43	Moderate boreal climate, moderately dry to fresh, strong, nutritious spruce-fir wood
2	2	7	Deciduous	44	Suboceanic, moderately dry to fresh, nutrient-rich beech forest
2	4	5	Deciduous	45	Suboceanic, moist, fast-growing, nutrient-rich sycamore and ash forest of the montane stage
2	2	5	Deciduous	46	Suboceanic, moderately dry to fresh, vigorous, nutrient-rich oak forest
2	4	5	Deciduous	47	Suboceanic, moist, vigorous, nutritious beech forest
2	4	7	Deciduous	48	Suboceanic, moist, nutrient-rich oak-hornbeam-ash wood
2	5	7	Deciduous	49	Suboceanic, moist, nutrient-rich black alder floodplain forest
2	7	5	Deciduous	50	Suboceanic, moist, flooded, vigorous, nutrient-rich elm and English oak floodplain forest
2	2	7	Deciduous	51	Suboceanic moderately dry to fresh, nutrient-rich oak wood alternating with dry wood

(continued)

Table 3.4 (continued)

Ecocli-matic zones	Condition of the soil moisture	Type of nutrient cycle	Forest type (CORINE Land Cover)	ID	Class of the ecosystem type forest/wood
3	2	7	Deciduous	52	Central European to subcontinental, moderately dry to fresh, nutrient-rich winter lime wood
1	2	7	Deciduous	53	Temperate boreal climate, moderately dry to fresh, nutrient-rich sycamore beech forest
2	2	7	Deciduous	54	Suboceanic, moderately dry to fresh, nutrient-rich hornbeam-beech forest
3	2	7	Deciduous	55	Central European to subcontinental, moderately dry to fresh, nutrient-rich hornbeam forests
2	2	7	Deciduous	56	Suboceanic, moderately dry to fresh, nutrient-rich beech forest
2	2	7	Deciduous	57	Suboceanic, moderately dry to fresh, nutrient-rich mountain elm-summer linden-blockwood trees
2	4	7	Deciduous	58	Suboceanic, moist, nutrient-rich beech forest
2	4	7	Deciduous	59	Suboceanic, moist, nutrient-rich ash wood
2	7	7	Deciduous	60	Suboceanic, moist, flooded, nutrient-rich floodplain forest with Salix x rubens
2	1	8	Deciduous	61	Suboceanic, dry, nutrient-rich, carbonate-rich sessile oak dry forest
2	1	8	Deciduous	62	Suboceanic, dry, nutrient-rich, carbonate-containing oak wood
2	2	8	Deciduous	63	Suboceanic, moderately dry to fresh, nutrient-rich, carbonate-rich sunny slope beech forest
2	2	8	Deciduous	64	Suboceanic, moderately dry to fresh, nutrient-rich, carbonate-rich beech wood/forest stand
2	4	8	Deciduous	65	Suboceanic, moist, nutrient-rich, carbonate-containing ash wood
1	8	9	Deciduous	66	Moderate boreal climate, very wet, nutrient-poor organic raised bog
2	8	9	Deciduous	67	Suboceanic, very moist, nutrient-poor organic raised bog

(continued)

Table 3.4 (continued)

Ecocli-matic zones	Condition of the soil moisture	Type of nutrient cycle	Forest type (CORINE Land Cover)	ID	Class of the ecosystem type forest/wood
1	5	10	Deciduous	68	Moderate boreal climate, moist, fairly nutrient-poor organic raised bog wood
1	5	10	Deciduous	69	Moderate boreal climate, humid, fairly nutrient-poor organic Carpathian birch forest
3	5	10	Deciduous	70	Central European to subcontinental, moist, fairly nutrient-poor organic bog birch wood
2	4	11	Deciduous	71	Suboceanic, moist, moderately nutritious organic black alder wood
2	5	11	Deciduous	72	Suboceanic, moist, moderately nutritious organic black alder wood
2	5	12	Deciduous	73	Suboceanic, moist, strong, nutritious organic black alder wood
2	7	12	Deciduous	74	Suboceanic, moist flooded, vigorous, nutrient-rich organic grey alder forest
2	7	12	Deciduous	75	Suboceanic, moist flooded, vigorous, nutritious organic Prunus padus ash wood
1	2	14	Conifer	76	Moderate boreal climate, moderately dry to fresh, quite nutrient-poor carbonate-containing mountain pine krummholz
1	2	14	Conifer	77	Temperate boreal climate, moderately dry to fresh, fairly nutrient-poor, carbonate-rich spruce wood
1	2	14	Mixed	78	Temperate boreal climate, moderately dry to fresh, fairly nutrient-poor carbonate-rich spruce-beech wood

3.1.2.1 Data Sources for Mapping the Potential Habitat Service of Ecosystem Type Classes in Germany

The data basis for the assessment of habitat function was derived from the BERN model (Schlutow et al. 2024., see https://github.com/bern-model/BERN). The BERN model (Bioindication for Ecosystem Regeneration towards Natural conditions) was developed to integrate ecological cause-effect relationships into studies on environmental status assessment and prediction. Qualitative knowledge on the relationship between site types and vegetation communities is widely available in the extensive phytosociological publications. The BERN database documents

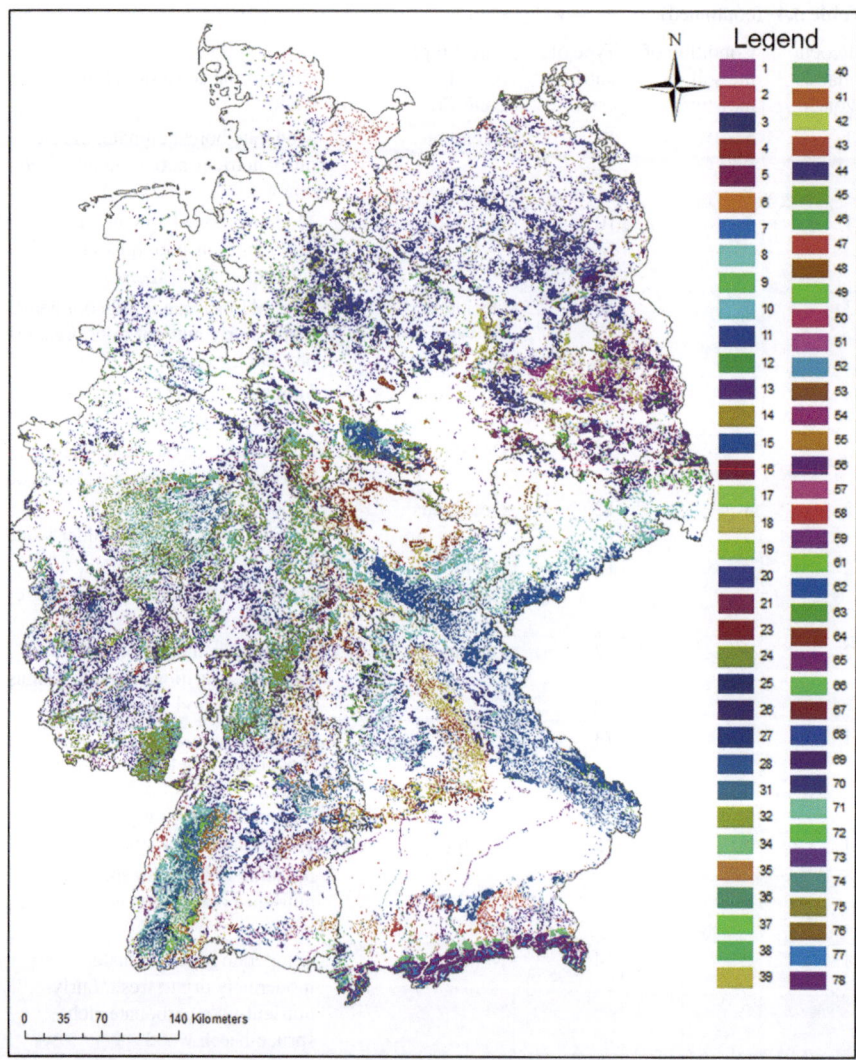

Fig. 3.1 Distribution of ecosystem type classes in Germany (for an explanation of the class code, see Table 3.4 above) (Author's own illustration)

reference data on typical site parameters for the occurrence of plant communities in which their diagnostic species are in competitive equilibrium with each other and in homeostatic equilibrium with the site factors. The BERN model makes part of this knowledge available in numerical form for computer-aided ecosystem modelling. The BERN model combines the niches realised by species, which mainly form the competitive stable structure of a natural plant community, to determine the basic multifactorial niche of this community. The database currently defines the niche of 2412 European plant species for the soil properties pH, base saturation, carbon-nitrogen ratio and wetness index as well as the climatic properties continentality, length of the growing season, climatic water balance and solar radiation (PAR) by analysing currently about 63,700 vegetation-relevant with site information (Fig. 3.2).

The BERN database contains 975 European plant communities and links to the diagnostic species of the communities. The results of BERN modelling are intended to promote or initiate the development of vegetation types that are highly resilient to site changes and can develop a high potential for ecosystem services.

For the modelling, data had to be collected to ensure that the self-organisation potential and thus the adaptability of the vegetation to changing site factors was represented in its original state. For this purpose, the spontaneously occurring plant communities had to be analysed from data collections that were recorded at largely unpolluted or already influenced sites where the site factors were still in equilibrium. Of particular interest are very early recordings, preferably those from before 1960.

There are only a few records with measurement data on abiotic site parameters from the period before the industrialisation wave in the second half of the twentieth century. And the available measured values from this period come from nonstandardised measurements.

However, the literature on vegetation synoptic tables generally also contains verbal information on soil, water and climate factors at the site. This information could be compared with existing databases of reference site forms according to their (then) definition and thus site parameters could be assigned to the locations of the municipalities by analogy. A detailed description of the data basis and methods can be found in Schlutow et al. (2024).

Only plant communities with clearly definable site constancy are included in the BERN database. While the ecological niches of the communities in the marginal

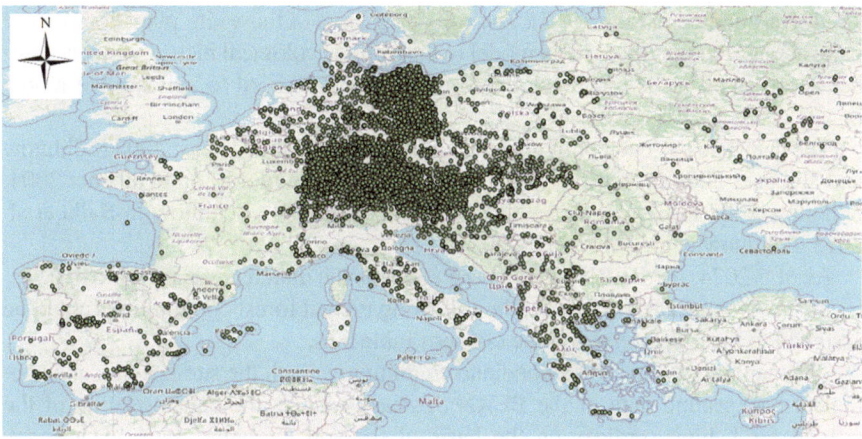

Fig. 3.2 Distribution of relevés in the BERN5.1 database. As of May 2024. (Sources: Authors' own presentation; OpenStreetMap)

areas (ecotones) can overlap considerably, the optimal areas are clearly delimited from each other.

Generally accessible publications of vegetation surveys in the relevant phytosociological literature were taken into account. Only synoptic tables were used, and only if they were accompanied by sufficient information on the site characteristics of the locations.

Only those forest, pasture (including dry and wet heaths), meadow (extensive grassland only), water body, bog and marsh communities that can be preserved in the long term (taking into account nature conservation management where appropriate) were included in the database.

The databases of the BERN model cover the entire area of Germany (Ellenberg 1996, Härdtle 1984, 1989, 1995a, b, Härdtle et al. 2004, Hartmann and Jahn 1967, Hofmann 1969, Hofmann and Pommer 2013, Hundt 1964, Hrivnák 2002, Issler 1924, 1926, 1942, Klapp 1954, 1965, Krausch 1962, Krieger 1937, Liebert 1988, Lohmeyer 1957, 1962, Mahn 1959, 1965, Matuszkiewicz and Matuszkiewicz 1956, Matuszkiewicz and Borowik 1958, Matuszkiewicz and Traczyk 1958,Matuszkiewicz 1962, Oberdorfer 1957, 1992–1998, 2001, Passarge 1960, 1964, Passarge and Hofmann 1968, Preising 1953, Preising et al. 1990a, b and Preising et al. 1997, Pott 1992, Pottgießer and Sommerhäuser 2004, von Rochow 1951, Schmidt et al. 2002, Schubert 1960, 1991, Schubert et al. 1995, Slobodda 1982, 1987, Succow 1974, 1988, Succow and Joosten 2001, Tüxen 1937, 1958, Tüxen and Westhoff 1963, Volk 1937, Willner 2002, Willner and Grabherr 2007, Wolfram 1996).

Outside Germany, a further 115 phytosociological literature sources were analysed and the corresponding data from the surveys were transferred to the BERN database.

The ecological niches of the biotic communities determined on the basis of the BÜK1000N and Level II datasets were compared with 280 site-plant community pairs from historical surveys (Klapp 1965; Krieger 1937; Mahn 1965; Schubert 1960; Succow 1988; Succow and Joosten 2001; Volk 1937; Wolfram 1996; Succow 1974; Hartmann and Jahn 1967). The soil parameters of these site-plant community pairs are not fed into the BERN model to determine ecological niches, as their number is too small to be representative. Instead, they are used to validate the model results of the BERN database.

The database has been built up and continuously expanded since 2002 (Schlutow and Hübener 2002 in Achermann and Bobbink 2003, Schlutow and Hübener 2004, Schlutow 2007 in De Vries et al. 2007, Nagel et al. 2010, Schlutow in Balla et al. 2013, Schlutow et al. 2015, 2018, 2024.

As of 5/2024, the BERN5.1 database contains the data records in Table 3.5.

Based on the BERN database, the following typical forest communities could be assigned to the 78 ecosystem type classes (Table 3.6).

The mapping of the forest/woodland communities with the site-specific and ecological background data of the BERN database enables the categorisation of the criteria for habitat services on the basis of the soil overview map of Germany (BÜK1000N), combined with the climate grid maps of the DWD (2012) and the Corine Land Cover (UBA 2015) as follows:

Table 3.5   Number of data records in the BERN5.1 database (as at 5/2024)

Stand	BERN5.1 (2024)
Number of vegetation data (Europe) summarised in synoptic tables, with description of typical site factors	63,760
of which vegetation-relevant biotic communities that occur in Germany	25,954
Vegetation-relevant with site-specific measured values of soil chemistry (Europe-wide)	965
of which Level II locations in Germany	85
of which Level II locations in other European countries	600
of which other vegetation surveys with soil chemistry data in Germany	280
Number of reference soil profiles from the 1:1 million soil overview map of Germany with site-typical soil chemistry measurements that were assigned to the plant communities identified as typical in the BERN database by expert reports.	674
Plant species with species-specific ecological niches for 8 site parameters	2412
Plant communities with derived non-linear ecological niches for 8 site parameters	975
Of which natural wood communities	462

(a) Hemerobicity (degree of naturalness)

The assignment of the forest/woodland communities to the ecosystem type classes (Table 3.6) enables the assessment of the hemerobia status in each ecosystem type class. A Germany-wide survey of the degree of naturalness (degree of hemerobicity) by Walz and Stein (2014) was used to validate the assessment. The assessment of hemerobicity is based on the following estimates.

Primeval forests that have never been cleared or afforested by humans are considered natural. In Germany, such forests can only be found in the form of high mountain dwarf pine forests in the transition area to the forest-free high alpine zone. The treeless alpine mat and rock vegetation types on non-forestable sites are also considered natural. Non-forested wet sites in the riparian zone of water bodies also form natural floodplain meadows, reedbeds and wet meadows.

Natural ecosystem type classes are forests that were originally afforested but have a species structure that largely corresponds to the potential natural forest community, such as alluvial, bog and wetland forests, site-typical mixed deciduous forests (e.g., coastal dune-sand pine forest, spruce-fir mountain forest, winter lime-hornbeam forest, carbonate beech forest).

Near-natural with significant impairments are ecosystem type classes whose vegetation type (still) corresponds to today's potentially natural woody community, but the site parameters have already changed so much, e.g., due to acidification, eutrophication, drainage, etc., that they no longer offer the vegetation type an optimal opportunity to exist.

Non-natural forests are forests whose tree species composition is severely impoverished compared to today's potentially natural forest community and/or is not typical for the site. Non-natural forests with non-native tree species can also lead to a change in site characteristics (e.g., coniferous forests on naturally medium soils lead to soil acidification).

Table 3.6 Current forest/woodland communities, potential woodland communities and number of their characteristic plant species and associated animal species, which the BERN model assigns to the ecosystem type classes (IDs in Table 3.4)

Class ID	Name of the ecosystem type class (Table 3.4)	Contemporary forest/ timber communities (Current status)	Number of characteristic plant species	Number of ecosystem-typical animal species protected under EU law	Potential forest community (pristine state)	Number of characteristic plant species	Number of ecosystem-typical animal species protected under EU law
1	Central European to subcontinental, dry, nutrient-poor pine forest	Dicrano-Cultopinetum HOFMANN 2002	14	59	Cladonio-Pinetum sylvestris typicum PASS. 1956	28	60
2	Suboceanic, dry, nutrient-poor pine forest	Empetro nigri-Pinetum sylvestris LIBB. et SISS. 1939	23	38	Empetro nigri-Pinetum sylvestris LIBB. et SISS. 1939	23	38
3	Central European to subcontinental, dry, nutrient-poor pine wood	Cladonio-Pinetum sylvestris typicum PASS. 1956	28	60	Cladonio-Pinetum sylvestris typicum PASS. 1956	28	60
4	Suboceanic, moderately dry to fresh, nutrient-poor pine forest	Vaccinio myrtilli-Kultopinetum HOFMANN 2002	17	59	Dicrano-Fagetum sylvatici PASS. et HOFM. 1965	10	99
5	Suboceanic, moderately dry to fresh, poor larch forest	Myrtyllo-Cultolaricetum deciduae HOFMANN et POMMER 2013	6	?	Dicrano-Fagetum sylvatici PASS. et HOFM. 1965	10	99
6	Suboceanic, moderately dry to fresh, nutrient-poor beech forest / woodland	Dicrano-Fagetum sylvatici PASS. et HOFM. 1965	10	99	Dicrano-Fagetum sylvatici PASS. et HOFM. 1965	10	99

7	Suboceanic, moderately dry to fresh, nutrient-poor spruce wood	Calamagrostio villosae-Piceetum typicum VOLK 1939	24	92	Calamagrostio villosae-Piceetum typicum VOLK 1939	24	92
8	Suboceanic, moderately dry to fresh, nutrient-poor spruce forest	Dicrano-Cultopiceetum SCHUBERT 1972	6	92	Dicrano-Fagetum sylvatici PASS. et HOFM. 1965	10	99
9	Suboceanic, moist, nutrient-poor pine forest	Molinio-Cultopinetum HOFMANN 1964	8	12	Betulo-Quercetum roboris molinetosum (TX. 1937) SCAMONI et PASSARGE 1959	24	100
10	Suboceanic, moist, fairly nutrient-poor pine forest	Pteridio-Cultopinetum HOFMANN 1964	19	12	Vaccinio myrtilli-Fagetum sylvatici typicum PASS. 1965	15	98
11	Central European to subcontinental, moist, nutrient-poor spruce wood	Vaccinio uliginosi-Piceetum HARTM. 1953	17	92	Vaccinio uliginosi-Piceetum HARTM. 1953	17	92
12	Central European to subcontinental, dry, fairly nutrient-poor sessile oak wood	Calamagrostio arundinaceae-Quercetum petraeae SCAMONI et PASSARGE 1959	34	100	Calamagrostio arundinaceae-Quercetum petraeae SCAMONI et PASSARGE 1959	34	100

(continued)

Table 3.6 (continued)

Class ID	Name of the ecosystem type class (Table 3.4)	Contemporary forest/timber communities (Current status)	Number of characteristic plant species	Number of ecosystem-typical animal species protected under EU law	Potential forest community (pristine state)	Number of characteristic plant species	Number of ecosystem-typical animal species protected under EU law
13	Central European to subcontinental, dry, fairly nutrient-poor pine forest	Festuco-Cultopinetum HOFMANN 1964	14	59	Vaccinio vitis-ideae-Quercetum (roboris) OBERD. (1957) 1992	16	101
14	Suboceanic, moderately dry to fresh, fairly nutrient-poor beech forest/woodland	Vaccinio myrtilli-Fagetum sylvatici typicum PASS. 1965	15	98	Vaccinio myrtilli-Fagetum sylvatici typicum PASS. 1965	15	98
15	Moderate boreal climate, moderately dry to fresh, quite nutrient-poor beech wood/forest	Calamagrostio arundinaceae-Abieto-Fagetum sylvatici dryopteritetosum HARTM. et JAHN 1967	18	101	Calamagrostio arundinaceae-Abieto-Fagetum sylvatici dryopteritetosum HARTM. et JAHN 1967	18	101
16	Suboceanic, moist, fairly nutrient-poor English oak forest	Betulo-Quercetum roboris molinietosum (TX. 1937) SCAMONI et PASSARGE 1959	24	100	Betulo-Quercetum roboris molinietosum (TX. 1937) SCAMONI et PASSARGE 1959	24	100
17	Suboceanic, moist, fairly nutrient-poor fir forest	Vaccinio-Abietetum oxalietosum OBERD. 1957	15	92	Vaccinio-Abietetum oxalietosum OBERD. 1957	15	92
18	Central European to subcontinental, dry, moderately nutrient-rich pine forest	Calamagrostio-Cultopinetum HOFMANN 2002	11	12	Potentillo albae-Quercetum petraeae-roboris LIBBERT 1933	37	101

19	Central European to subcontinental, dry, moderately nutritious sessile oak wood	Potentillo albae-Quercetum petraeae-roboris LIBBERT 1933	37	101	Potentillo albae-Quercetum petraeae-roboris LIBBERT 1933	37	101
20	Suboceanic, moderately dry to fresh, moderately nutrient-rich pine forest	Rubo-Cultopinetum HOFMANN 2002	15	12	Maianthemo-Fagetum sylvatici PASS. 1959	14	98
21	Central European to subcontinental, moderately dry to fresh, moderately nutrient-rich pine forest	Calamagrostio-Myrtillo-Cultopinetum HOFMANN and Pommer 2013	24	12	Agrostio-Quercetum roboris deschampsietosum flexuosaea PASS. 1953 em. SCHUB. 1995	24	101
22	Suboceanic, moist, moderately nutrient-rich oak forest	Dryopteri-Cultoquercetum HOFMANN 2002	12	1	Holco mollis-Quercetum (robori-petraeae) LEMÉE 1937 corr. et em. OBERD. 1992	10	101
23	Suboceanic, dry, moderately nutrient-rich, carbonate-rich pine forests of the Alpine valleys	Erico carneae-Pinetum sylvestris BR.-BL. in BR.-BL. et al. 1939	38	2	Erico carneae-Pinetum sylvestris BR.-BL. in BR.-BL. et al. 1939	38	2
24	Suboceanic, moderately dry to fresh, moderately nutrient-rich beech forest/wood	Maianthemo-Fagetum sylvatici PASS. 1959	14	98	Maianthemo-Fagetum sylvatici PASS. 1959	14	98

(continued)

Table 3.6 (continued)

Class ID	Name of the ecosystem type class (Table 3.4)	Contemporary forest/timber communities (Current status)	Number of characteristic plant species	Number of ecosystem-typical animal species protected under EU law	Potential forest community (pristine state)	Number of characteristic plant species	Number of ecosystem-typical animal species protected under EU law
25	Central European to subcontinental, moderately dry to fresh, moderately nutritious Sessile oak-beech wood	Holco mollis-Quercetum (robori-petraeae) LEMÉE 1937 corr. et em. OBERD. 1992	10	101	Holco mollis-Quercetum (robori-petraeae) LEMÉE 1937 corr. et em. OBERD. 1992	10	101
26	Central European to subcontinental, moderately dry to fresh, moderately nutrient-rich lime-hornbeam wood	Bromo-Carpinetum betuli primulaetosum HOFM. 1968	30	115	Bromo-Carpinetum betuli primulaetosum HOFM. 1968	30	115
27	Moderate boreal climate, moderately dry to fresh, moderately nutrient-rich beech wood/forest stand	Luzulo luzuloides-Fagetum sylvatici typicum MEUSEL 1937	12	98	Luzulo luzuloides-Fagetum sylvatici typicum MEUSEL 1937	12	98
28	Suboceanic, moderately dry to fresh, moderately nutrient-rich spruce forest	Brachypodio sylvaticae-Cultopiceetum SCHUBERT 1972	9	13	Maianthemo-Fagetum sylvatici typicum PASS. 1959	14	98

#	Description						
29	Suboceanic, moderately dry to fresh, moderately nutritious spruce-fir wood	Luzulo-Abietetum typicum OBERD. 1957	22	92	Luzulo-Abietetum typicum OBERD. 1957	22	92
30	Suboceanic, moderately dry to fresh, moderately nutrient-rich Douglas fir forest	Rubo-Deschampsio-Cultodouglasietum menziesii HOFMANN et POMMER 2013	7	?	Maianthemo-Fagetum sylvatici typicum PASS. 1959	14	98
31	Suboceanic, moderately dry to fresh, moderately nutrient-rich fir-beech forest	Luzulo-Abieto-Fagetum typicum HARTM. et JAHN 1967	16	100	Luzulo-Abieto-Fagetum typicum HARTM. et JAHN 1967	16	100
32	Suboceanic, moderately dry to fresh, moderately nutritious English oak - beech wood	Carici piluliferae-Fagetum sylvatici agrostidetosum PASS. 1956	18	98	Carici piluliferae-Fagetum sylvatici agrostidetosum PASS. 1956	18	98
33	Suboceanic, moderately dry to fresh, moderately nutrient-rich oak forest	Poo-Cultoquercetum HOFMANN 2002	15	1	Maianthemo-Fagetum sylvatici typicum PASS. 1959	14	98
34	Moderate boreal climate, moist, moderately nutrient-rich Fir-beech wood	Equiseto sylvatici-Abietetum albae MOOR 1952	20	92	Equiseto sylvatici-Abietetum albae MOOR 1952	20	92

(continued)

Table 3.6 (continued)

Class ID	Name of the ecosystem type class (Table 3.4)	Contemporary forest/timber communities (Current status)	Number of characteristic plant species	Number of ecosystem-typical animal species protected under EU law	Potential forest community (pristine state)	Number of characteristic plant species	Number of ecosystem-typical animal species protected under EU law
35	Suboceanic, moist, moderately nutrient-rich fir-beech forest	Luzulo-Abieto-Fagetum sylvatici dryopteridetosum HARTM. et JAHN 1967	14	101	Luzulo-Abieto-Fagetum sylvatici dryopteridetosum HARTM. et JAHN 1967	14	101
36	Suboceanic, moist, moderately nutrient-rich oak-hornbeam forest	Querco roboris-Carpinetum betuli TX. 1937 primulaetosum	17	97	Querco roboris-Carpinetum betuli TX. 1937 primulaetosum	17	97
37	Suboceanic, moist, moderately nutritious fir wood	Luzulo-Abietetum athyrietosum OBERD. 1957	21	92	Luzulo-Abietetum athyrietosum OBERD. 1957	21	92
38	Suboceanic, dry, moderately nutrient-rich, carbonate pine forest	Brachypodio pinnati-Cultopinetum HOFMANN 2002	24	97	Cephalanthero-Pinetum sylvestris ELLENBERG et KLÖTZLI 1972	26	99
39	Suboceanic, moderately dry to fresh, moderately nutrient-rich carbonate spruce forest	Calamagrostio variae-Piceetum SCHWEINGRUBER 1972	11	92	Carici-Fagetum sylvatici typicum MOOR 1952 em. LOHM 1953	21	104
40	Central European to subcontinental, dry, vigorous, nutrient-rich rock maple - sessile oak wood	Aceri monspessulani-Quercetum petraeae OBERD. 1957	26	35	Aceri monspessulani-Quercetum petraeae OBERD. 1957	26	35

41	Central European to subcontinental, moderately dry to fresh, vigorous, nutrient-rich hornbeam forest	Galio-Carpinetum betuli typicum OBERDORFER 1957	19	115	Galio-Carpinetum betuli typicum OBERDORFER 1957	19	115
42	Central European to subcontinental, moderately dry to fresh, vigorous, nutrient-rich winter lime wood	Luzulo luzuloides-Tilietum cordatae GRABHERR et MUCINA 1989	21	100	Luzulo luzuloides-Tilietum cordatae GRABHERR et MUCINA 1989	21	100
43	Temperate boreal climate, moderately dry to fresh, strong, nutritious spruce-fir wood	Dentario bulbiferae-Fagetum sylvatici typicum LOHMEYER 1962	9	96	Dentario bulbiferae-Fagetum sylvatici typicum LOHMEYER 1962	9	96
44	Suboceanic, moderately dry to fresh, vigorous, nutritious beech forest	Asperulo-Fagetum sylvatici typicum MAYER 1964	25	95	Asperulo-Fagetum sylvatici typicum MAYER 1964	25	95
45	Suboceanic, moist, fast-growing, nutrient-rich sycamore and ash forest of the montane stage	Adoxo-Aceretum pseudoplatani PASSARGE 1959	40	93	Adoxo-Aceretum pseudoplatani PASSARGE 1959	40	93
46	Suboceanic, moderately dry to fresh, vigorous, nutrient-rich oak forest	Asperulo-Cultoquercetum HOFMANN 2002	17	1	Asperulo-Fagetum sylvatici typicum MAYER 1964	25	95

(continued)

Table 3.6 (continued)

Class ID	Name of the ecosystem type class (Table 3.4)	Contemporary forest/timber communities (Current status)	Number of characteristic plant species	Number of ecosystem-typical animal species protected under EU law	Potential forest community (pristine state)	Number of characteristic plant species	Number of ecosystem-typical animal species protected under EU law
47	Suboceanic, moist, vigorous, nutritious beech forest	Asperulo-Fagetum sylvatici dryopteridetosum SCAM. 1967	31	95	Asperulo-Fagetum sylvatici dryopteridetosum SCAM. 1967	31	95
48	Suboceanic, moist, strong, nutritious oak-hornbeam-ash wood	Stellario holosteae-Carpinetum betuli HARTMANN 1959	11	97	Stellario holosteae-Carpinetum betuli HARTMANN 1959	11	97
49	Suboceanic, moist, vigorous, nutrient-rich black alder floodplain forest	Stellario-Alnetum typicum LOHMEYER 1957	24	120	Stellario-Alnetum typicum LOHMEYER 1957	24	120
50	Suboceanic, moist, flooded, vigorous, nutrient-rich elm and English oak floodplain forest	Querco-Ulmetum ISSLER 1953	13	119	Querco-Ulmetum ISSLER 1953	13	119
51	Suboceanic moderately dry to fresh, nutrient-rich oak wood alternating with dry wood	Sambuco-Quercetum roboris HOFMANN 1965	28	101	Sambuco-Quercetum roboris HOFMANN 1965	28	101

52	Central European to subcontinental, moderately dry to fresh, nutrient-rich winter lime wood	Carici albae-Tilietum cordatae MÜLLER et GÖRS 1958	15	94	Carici albae-Tilietum cordatae MÜLLER et GÖRS 1958	15	94
53	Temperate boreal climate, moderately dry to fresh, nutrient-rich sycamore beech forest	Aceri pseudoplatani-Fagetum sylvatici BARTSCH et BARTSCH 1940	14	19	Aceri pseudoplatani-Fagetum sylvatici BARTSCH et BARTSCH 1940	14	19
54	Suboceanic, moderately dry to fresh, nutrient-rich hornbeam-beech forest	Aceri platanoides-Carpinetum betuli KLIKA 1941	23	93	Aceri platanoides-Carpinetum betuli KLIKA 1941	23	93
55	Central European to subcontinental, moderately dry to fresh, nutrient-rich hornbeam-beech forest	Mercuriali-Carpinetum betuli typicum SCAM. et PASS. 1959	18	95	Mercuriali-Carpinetum betuli typicum SCAM. et PASS. 1959	18	95
56	Moderate boreal climate, moderately dry to fresh, nutrient-rich Beech wood/forest	Seslerio variae-Fagetum sylvatici KNAPP 1948	24	103	Seslerio variae-Fagetum sylvatici KNAPP 1948	24	103
57	Suboceanic, moderately dry to fresh, nutrient-rich mountain elm-summer linden-blockwood trees	Aceri platanoides-Tilietum platyphylli WINTERHOFF 1962	25	93	Aceri platanoides-Tilietum platyphylli WINTERHOFF 1962	25	93
58	Suboceanic, moist, nutrient-rich beech forest	Hordelymo-Fagetum sylvatici TX. 1937 dryopteridetosum	14	95	Hordelymo-Fagetum sylvatici TX. 1937 dryopteridetosum	14	95

(continued)

Table 3.6 (continued)

Class ID	Name of the ecosystem type class (Table 3.4)	Contemporary forest/timber communities (Current status)	Number of characteristic plant species	Number of ecosystem-typical animal species protected under EU law	Potential forest community (pristine state)	Number of characteristic plant species	Number of ecosystem-typical animal species protected under EU law
59	Suboceanic, moist, nutrient-rich ash wood	Ribo sylvestris-Fraxinetum LEMÉE 1937 corr. PASSARGE 1958	7	112	Ribo sylvestris-Fraxinetum LEMÉE 1937 corr. PASSARGE 1958	7	112
60	Suboceanic, moist, flooded, nutrient-rich floodplain forest with Salix x rubens	Salicetum albae ISSLER 1926	18	119	Salicetum albae ISSLER 1926	18	119
61	Suboceanic, dry, nutrient-rich, carbonate-rich sessile oak rocky dry forest	Buxo-Quercetum pubescentis BR.-BL. 1932	27	35	Buxo-Quercetum pubescentis BR.-BL. 1932	27	35
62	Suboceanic, dry, nutrient-rich, carbonate-containing oak wood	Cytiso nigricantis-Quercetum roboris OBERD. 1957	26	101	Cytiso nigricantis-Quercetum roboris OBERD. 1957	26	101
63	Suboceanic, moderately dry to fresh, nutrient-rich, carbonate-rich sunny slope Beech wood/forest	Cephalanthero-Fagetum sylvatici typicum OBERDORFER 1957	47	104	Cephalanthero-Fagetum sylvatici typicum OBERDORFER 1957	47	104

64	Suboceanic, moderately dry to fresh, nutrient-rich, carbonate-rich Beech wood/forest	Carici-Fagetum sylvatici typicum MOOR 1952 em. LOHM 1953	21	104	Carici-Fagetum sylvatici typicum MOOR 1952 em. LOHM 1953	21	104
65	Suboceanic, moist, nutrient-rich, carbonate-containing ash wood	Carici remotae-Fraxinetum excelsi W. KOCH 1926 ex FAB. 1936	51	113	Carici remotae-Fraxinetum excelsi W. KOCH 1926 ex FAB. 1936	51	113
66	Moderate boreal climate, very wet, nutrient-poor organic raised bog	Erico-Sphagnetum magellanici (OSVALD 1923) MOORE 1968	20	48	Erico-Sphagnetum magellanici (OSVALD 1923) MOORE 1968	20	48
67	Suboceanic, very wet, nutrient-poor organic raised bog	Ledo-Sphagnetum magellanici SUKOPP ex NEUHÄUSL 1969	11	47	Ledo-Sphagnetum magellanici SUKOPP ex NEUHÄUSL 1969	11	47
68	Moderate boreal climate, moist, fairly nutrient-poor organic raised bog wood	Sphagno-Piceetum KUOCH 1954	27	113	Sphagno-Piceetum KUOCH 1954	27	113
69	Moderate boreal climate, humid, fairly nutrient-poor organic Carpathian birch forest	Betulo carpaticae-Piceetum STÖCK. 1967	17	92	Betulo carpaticae-Piceetum STÖCK. 1967	17	92
70	Central European to subcontinental, moist, fairly nutrient-poor organic bog birch wood	Sphagno-Betuletum pubescentis DOING 1962	24	111	Sphagno-Betuletum pubescentis DOING 1962	24	111

(continued)

Table 3.6 (continued)

Class ID	Name of the ecosystem type class (Table 3.4)	Contemporary forest/timber communities (Current status)	Number of characteristic plant species	Number of ecosystem-typical animal species protected under EU law	Potential forest community (pristine state)	Number of characteristic plant species	Number of ecosystem-typical animal species protected under EU law
71	Suboceanic, moist, moderately nutrient-rich organic Black alder wood	Athyrio-Alnetum glutinosae TX. 1943	27	113	Athyrio-Alnetum glutinosae TX. 1943	27	113
72	Suboceanic, moist, moderately nutrient-rich organic Black alder wood	Carici elongatae-Alnetum glutinosae BODEUX 1955	17	112	Carici elongatae-Alnetum glutinosae BODEUX 1955	17	112
73	Suboceanic, moist, vigorous, nutritious, organic Black alder wood	Stellario-Alnetum typicum LOHMEYER 1957	24	120	Stellario-Alnetum typicum LOHMEYER 1957	24	120
74	Suboceanic, moist flooded, strong, nutritious organic grey alder wood	Alnetum incanae LÜDI 1921	20	120	Alnetum incanae LÜDI 1921	20	120
75	Suboceanic, moist flooded, vigorous, nutritious organic Prunus padus ash wood	Pruno-Fraxinetum excelsi OBERDORFER 1953	19	112	Pruno-Fraxinetum excelsi OBERDORFER 1953	19	112

76	Temperate boreal climate, moderately dry to fresh, fairly nutrient-poor carbonate-containing mountain pine krummholz	Erico carnae-Pinetum uncinatae BR.-BL. in BR.-BL. et al. 1939	16	5	Erico carnae-Pinetum uncinatae BR.-BL. in BR.-BL. et al. 1939	16	5
77	Temperate boreal climate, moderately dry to fresh, fairly nutrient-poor, carbonate-rich spruce wood	Adenostylo glabrae-Piceetum MAYER 1969	42	92	Adenostylo glabrae-Piceetum MAYER 1969	42	92
78	Temperate boreal climate, moderately dry to fresh, fairly nutrient-poor carbonate-rich spruce-beech wood	Lonicero alpiginae-Abieti-Fagetum sylvatici typicum OBERDORFER 1957	33	97	Lonicero alpiginae-Abieti-Fagetum sylvatici typicum OBERDORFER 1957	33	97

(b) Compositional completeness

The survey of the number of species is based on the classification of pristine natural vegetation (Table 3.6) to the site types from the intersection of the BÜK1000N (BGR 2014), the CORINE Land Use Map 2012 (UBA 2015), the Ecoclimatic Zone Mapping (Balla et al. 2013) and the BERN5 database of plant communities, their preferred locations and their characteristic species in Europe (Schlutow et al. 2024; Fig. 3.3).

The difference between the original state and the current state in terms of the number of characteristic species leads to the categorisation of current biodiversity in terms of plant species.

(c) Habitat value for fauna

The suitability of the ecosystem type classes as reproduction, feeding and refuge habitats for mammals, birds, insects, reptiles and amphibians could be categorised as the difference between the original state and the actual state in terms of the number of associated species. The description of the original states of the ecosystem type classes made it possible to estimate the number of ecosystem-typical animal species using the BERN database based on knowledge of the habitat requirements of indicator species according to Schlutow and Schröder (2021) (Table 3.6).

(d) Vulnerability/need for protection

The period in which the ecosystem type classes could be restored after destruction is shown in Table 3.7. The information on the ecosystem type classes includes

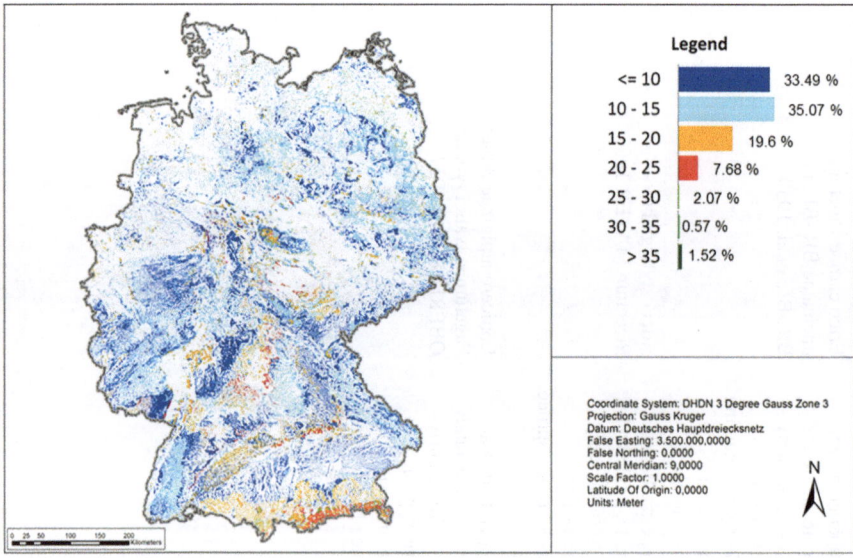

Fig. 3.3 Number of diagnostic pristine species, determined on the basis of a soil- and climate-specific assignment of the pristine natural vegetation with the BERN model. (Source: Authors´ own depiction)

Table 3.7 Natural lifespan of tree species in Germany [years]

	Natural minimum service life	Maximum natural service life
Abies alba	500	600
Acer tataricum	50	70
Alnus glutinosa	100	120
Alnus incana	50	100
Betula pendula	100	120
Betula pubescens	100	120
Betula pubescens carpatica	30	50
Carpinus betulus	100	150
Fagus sylvatica	200	300
Fraxinus excelsior	250	300
Larix decidua	200	400
Picea abies	200	300
Pinus mugo rotundata	80	150
Pinus sylvestris hercynica	200	300
Pinus sylvestris sylvestris	200	500
Populus nigra	100	150
Quercus petraea	500	800
Quercus pubescens	500	800
Quercus robur	500	800
Salix alba	50	80
Salix fragilis	50	80
Tilia platyphyllos	900	1000
Ulmus carpinifolia	300	600
Ulmus glabra	400	500
Ulmus laevis	200	250
Acer campestre	120	150
Acer platanoides	120	150
Acer pseudoplatanus	400	500
Pinus mugo mugo	80	150
Black pine (Pinus nigra nigra)	150	200
Populus tremula	100	150
Sorbus aucuparia	50	100
Sorbus torminalis	50	100
Tilia cordata	700	800
Ulmus campestre	120	150
Pseudotsuga menziesii	400	700

Sources: LWF (2024)

the main tree species. Mature ecosystems with deadwood, i.e., forests in which the tree species reach their natural maximum age, have a particularly high habitat suitability. However, such forests do not (yet) exist in Central Europe. The period of recoverability therefore results from the natural lifespan of the most important tree species in Germany.

(e) Restorability/recoverability/replaceability of habitats

Selected specially and strictly protected indicator species and their preferred habitats were recorded in the BERN5 database (Schlutow et al. 2024). On this basis, it was possible to determine the potential occurrence of protected species in the ecosystem type classes and the affiliation of the ecosystem type classes to protected habitat types or biotopes.

(f) Maturity

Maturity level: The maturity level of the forests depends on the time of stand establishment, the management method and the tree species composition. There are no primeval forests in Germany, so a rating of 5 is out of the question. A score of 4 was awarded to deciduous and mixed forests whose current vegetation corresponds to the potential natural vegetation. A score of 3 was awarded to purely coniferous forests whose main tree species is typical of the site. They can develop into mixed forests through natural regeneration, as pure coniferous forests would not naturally exist in Germany. Pure coniferous forests on non-typical sites and deciduous forests of non-native species were given a score of 2. The scores 1 and 0 were not awarded here, as only forests were considered. Data for the validation of this method was available throughout Germany for the approx. 1800 areas of the Forest Soil Condition Inventory - BZE II (Wellbrock et al. 2017).

(g) Position within the biotope network

The position of the ecosystem type classes in the biotope network system could be determined by assigning the ecosystem type classes (Table 3.6) to the elements of a biotope network as follows.

Main connecting elements:

• All near-natural alluvial forests and wetland forests that accompany watercourses. Stepping stone ecosystems:

• -Other wetlands and swamp forests
• dry forests and scrubland
• Moist shrub landscapes and semi-arid grasslands and
• all other semi-natural forests

Ecosystem type classes with low significance for the biotope network

• Fresh meadow
• Non natural forests

Nitrogen inputs lead to changes in the species composition of the ground vegetation of forests when critical loads are exceeded (see Sect. 3.1.2.2, Schlutow and Scheuschner 2023). In the data set of the soil condition survey 2006–2008 (Wellbrock et al. 2017), a correlation between N pollution and composition of the herb layer is only recognisable for the data sets of the blueberry-spruce forest (sub-oceanic, moderately dry to fresh, nutrient-poor spruce forest/woodland). Here, both the number of nitrophytes and the number of species in the forest edges and clearings increase with the difference between N deposition and N uptake capacity of the forest ecosystems. However, the number of protected species does not change significantly, as the proportion of protected plant species in the total number (1–2 at best) is very low. In this respect, the influence of land-use change on habitat function outweighs the influence of N inputs and is therefore not analysed further here. No valid statements can currently be made about the change in habitat function for animals as a function of N inputs.

3.1.2.2   Data Sources for Mapping the Primary Biomass Productivity of Ecosystem Type Classes in Germany

The estimation of the primary biomass productivity of the ecosystem type classes is based on the spatial distribution of tree species in today's forests and woodlands in Germany as well as on data on soil properties and climate.

Soil parameters: The database of the use-differentiated soil overview map of Germany 1:1 million (BGR 2014) contains reference profiles for forest sites on 70 different soil types. For each soil horizon in the profile, information is provided on the soil type, effective exchange capacity and potential cation exchange capacity, usable field capacity, storage density, groundwater level and humus content. Based on the soil type, the pore fractions can be determined horizontally according to the soil mapping instructions (AG Boden 2005). The horizon designation provides information about the depth and the influence of the backwater as well as the humus-containing layers. This provides all the necessary data for determining the soil-specific net primary productivity for German forests (at a scale of 1:1 million).

Climate parameters: To estimate the influence of climate on net primary productivity, the necessary basic data such as the long-term average annual mean temperature and the annual precipitation sum as a 30-year average were obtained from the grid data set of the German Weather Service from 1991–2020 (DWD 2021). The length of the vegetation period is calculated from the number of days in the year $\geq 10\,°C$ mean daily temperature (DWD 2021).

The tree species-specific net primary productivity results from the main tree species of the ecosystem type class (Table 3.4) in combination with Tables 2.8 and 2.9. The result of the estimate of net primary productivity in all current forests and woodlands in Germany (Table 3.4) is shown in Fig. 3.4.

The assessment of the primary biomass productivity of the ecosystem type classes of in their potential original state was based on the distribution of the potential forest community in Germany (Table 3.6). The soil parameters and the climate parameters were determined in accordance with Sect. 2.3. However, the tree species and their specific net primary productivity are derived from the main tree species of the potential natural community (Fig. 3.13).

The assessment of the primary biomass productivity of ecosystem type classes in their current state under the long-term influence of nitrogen deposition was based on the spatial distribution of current forests and woodlands in Germany (Fig. 3.10). The soil parameters and the climate parameters were the same as for the ecosystem type classes in their original state according to Sect. 2.3.

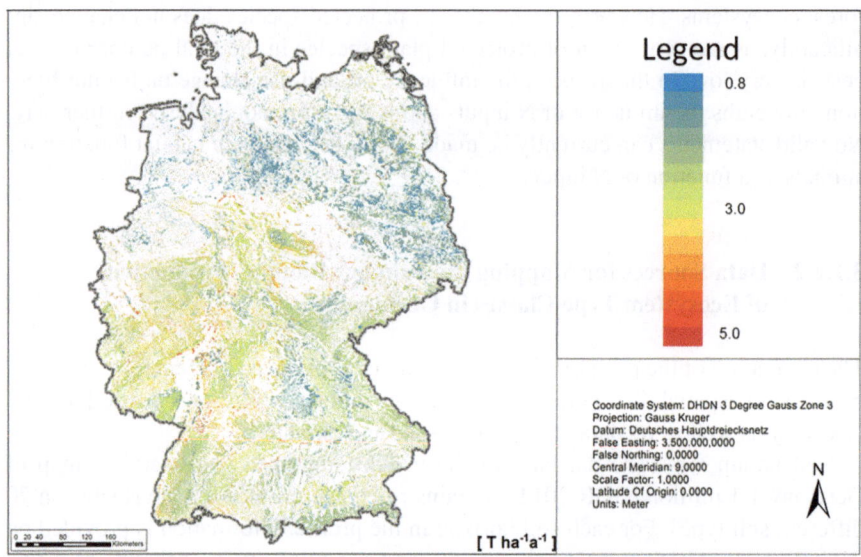

Fig. 3.4 Estimation of the potential biomass primary productivity of the ecosystem type classes on the basis of tree species-, soil- and climate-specific assessment criteria. (Source: Authors´ own depiction)

Since around 1985, N inputs have exceeded the harmless uptake rates in the soil and in the wood mass for a large part of the forest in Germany (Fig. 3.5). Although the proportion of overloaded forests (exceeding the ecosystem critical loads) has decreased since 1995, there has been no recognisable downward trend since 2015 (Schlutow and Scheuschner 2023; Schröder et al. 2023).

The persistently high nitrogen pollution in forest ecosystems affects all forests in the lowlands and in the low mountain ranges. Only the upper elevations of the low mountain ranges and the Alps are no longer too heavily polluted (Fig. 3.6).

The high nitrogen inputs in Germany since around 1975 initially compensated for the widespread nitrogen deficiency in the soil until then. As a result, the growth rates in wood mass for all tree species are now significantly higher than documented in the old yield tables from before 1975.

Therefore, the tree species-specific net primary productivity results from the main tree species of the ecosystem type class in combination with Table 3.8. This table contains growth rates of the wood of the tree species averaged over the entire rotation period, analysing current yield tables that already capture the previous influence of nitrogen deposition (Bösch 2001).

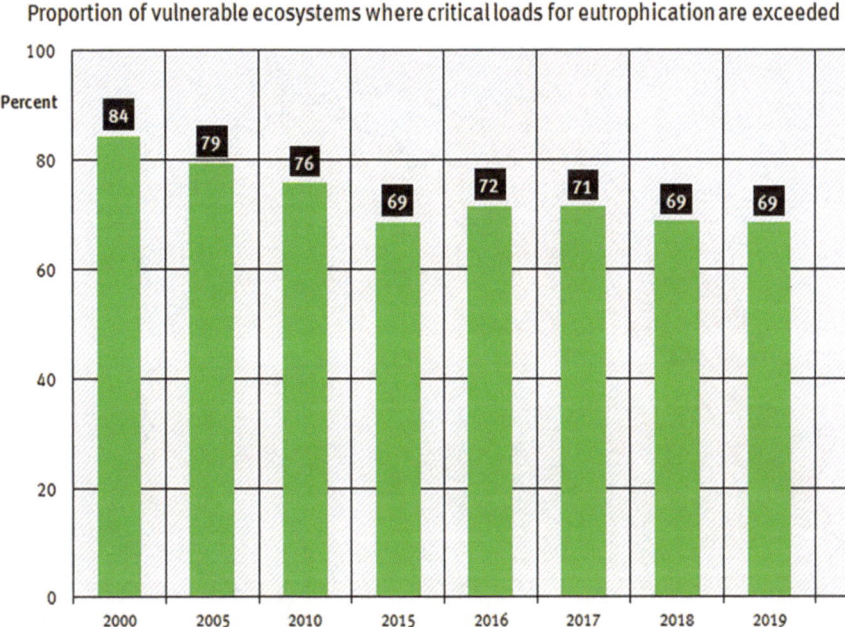

Fig. 3.5 Proportion of sensitive ecosystems in Germany in which the critical load for eutrophying nitrogen is exceeded. (Source: Kranenburg n.d.)

3.1.2.3   Data Sources for Mapping the Carbon Sequestration Capacity of Ecosystem Type Classes in Germany

(a) Clay content

The soil overview map of Germany BÜK1000N (BGR 2014) is the appropriate basis for a Germany-wide assessment of the influence of clay content on carbon storage capacity. The database of reference soil profiles does not contain any information on clay content. In the BÜK database, however, the soil type is indicated horizontally for the respective reference profile. The clay content could now be determined indirectly via the soil type of the horizon (Table 3.9) with the help of the pedological mapping guide KA5 (AG Boden 2005:142). The depth-weighted average of the real rooting depth resulting from the main tree type(s) was then calculated.

(b) Mass of litter and crop residues

Main tree species with information on species-specific litter mass and litter decomposition time: The BERN5 database contains an assignment of the European forest communities to the legend units of the BÜK1000N. This made it possible to link the ecosystem type classes with the mean litter mass production, decomposition time and other chemical parameters of the leaf litter according to Schachtschabel et al. (1998), Jacobsen et al. (2002), Scheffer and Ulrich (1960) and Madl (2017) (Table 3.10).

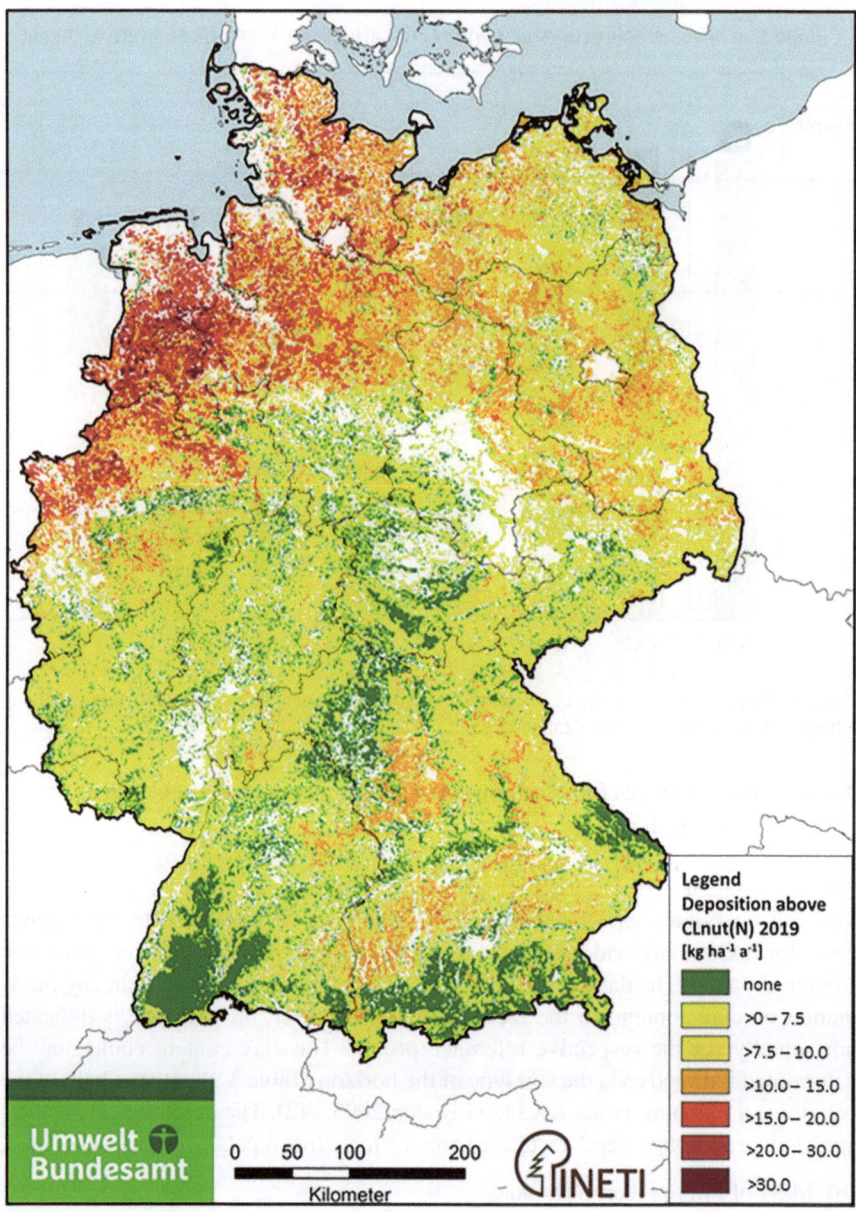

Fig. 3.6 Exceedance of the critical load for eutrophication ($CL_{nut}N$) due to atmospheric nitrogen deposition for 2019. (Source: Kranenburg n.d.)

Table 3.8 Intervals of net primary production (wood stage) of dominant and subdominant species in the actual state under the previous influence of nitrogen deposition (Bösch 2001)

Tree species	Average annual growth rates after 100 years [DGZ 100].	
	Net primary productivity of yield class I for logs with bark	Net primary productivity of the worst yield class for logs with bark
	$E_{max(Phyto)}$ [t dry matter ha^{-1} a]$^{-1}$	$E_{min(Phyto)}$ [t dry matter ha^{-1} a]$^{-1}$
Scots pine	5.8	2.3
Spruce	9.0	4.8
Douglas	10.0	4.7
Silver fir	9.8	5.4
European larch	5.8	2.2
Beech	9.5	5.8
Pedunculate and sessile oak	8.9	3.2
Alder	5.8	3.2
Birch, all species	6.2	2.6
Hornbeam	9.5	5.8
Ash	5.8	3.2
Lime, all species	7.1	4.2
Other deciduous tree species	6.6	2.8
Other conifer species	6.2	2.5

(c) Water content of the soil

Water content at field capacity: The database of the utilisation-differentiated soil overview map of Germany 1:1 million (BGR 2014) contains information on the usable field capacity for each soil horizon of the reference profile. Finally, the depth-weighted average of the real rooting depth resulting from the main tree species was calculated.

With the help of the classification according to Table 3.2 the range of volumetric water content can be assigned and evaluated for each ecosystem type class.

(d) pH value

The soil overview map of Germany 1:1 million (BGR 2014) contains information on pH levels for each soil horizon of the reference profile. According to the KA5 soil mapping guide, these levels can be categorised into pH ranges (AG Boden 2005, p. 367), from which the mean value was assigned to each horizon. The depth-level weighted mean value was then calculated for the actual rooting depth resulting from the main tree species(s).

(e) Ecoclimatic zones with average annual temperature and annual precipitation:

The ecosystem type class code contains the ecoclimatic zone (Table 3.1). This could be assigned to the ranges of annual mean temperature and annual mean precipitation in the long-term average 1991–2020 (DWD 2021). These ranges were

Table 3.9 Clay content of
soil types according to soil
mapping KA5 (AG Boden
2005, p. 142, Fig. 3.17)

Soil quality	Clay content [%]
Ss	3
Sl2	7
Sl3	10
Sl4	15
Slu	13
St2	11
St3	21
Su2	3
Su3	4
Su4	4
Ls2	21
Ls3	21
Ls4	21
Lt2	30
Lt3	40
Lts	35
Lu	24
Uu	4
Uls	13
We	4
Ut2	10
Ut3	14
Ut4	21
Tt	75
Tl	55
Tu2	52
Tu3	36
Tu4	28
Ts2	55
Ts3	40
Ts4	30

derived from the DWD grid data. The class boundaries used have a vegetation-ecological reference (Balla et al. 2013). For example, in the map of annual precipitation (Fig. 3.7), the red area corresponds to the main distribution area of the tree species oak and pine, the yellow-coloured area is dominated by beech, while green indicates the distribution area of beech and spruce and blue is more likely to be assigned to spruce.

An assessment of current C sequestration as a function of nitrogen inputs as part of the 2006–2008 soil status surveys *"could only be carried out to a limited extent, so that no reliable statements can be made in this regard"* (Wellbrock et al. 2017).

Table 3.10 Litter mass, decomposition time and other chemical parameters of leaf litter depending on the most important tree species (Schachtschabel et al. 1998; Jacobsen et al. 2002; Scheffer and Ulrich 1960; Madl 2017)

Tree species	Mass production of waste [t dry matter ha⁻¹ a⁻¹]	C/N ratio in fresh litter	pH value in fresh litter	Decomposition time (a) min	max
Ulmus spec.	5.161	28	6.5	0.4	1
Alnus glutinosa	5.534	15	5.6	0.5	1
Fraxinus excelsior	5.534	21	6.4	0.5	1
Robinia pseudacacia	5.534	14	5.4	0.5	1.5
Carpinus betulus	5.161	23	5.6	0.5	1.5
Castania sativa	5.161	23	4.5	0.5	1.8
Acer pseudoplatanus	5.161	52	4.5	1.5	2
Tilia cordata, T. platyphyllos	5.161	37	5.4	1.5	2
Quercus robur, Q. petraea	5.534	47	4.7	1.5	2.5
Betula pendula	5.161	50	5.5	1.5	2.5
Populus tremula	5.161	63	5.7	1.5	2.8
Quercus rubra	5.534	53	4.8	1.5	3
Fagus sylvatica	5.161	51	4.3	1.5	3
Picea abies	3.805	48	4.1	2.5	3.3
Pinus sylvestris	1.408	66	4.2	2.5	3.8
Pseudotsuga menziesii	3.347	77	4.3	2.5	3.8
Larix decidua	3.266	113	5.7	2.8	5

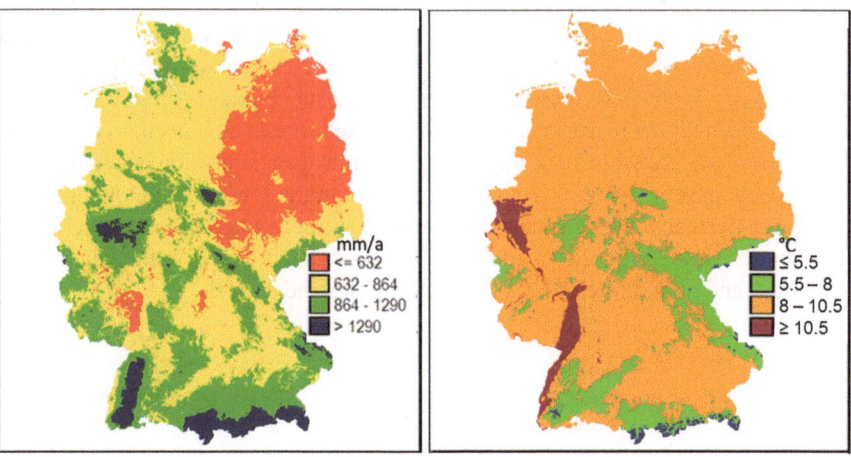

Fig. 3.7 Climate data for the long-term average 1981–2010 (DWD 2012), left: annual precipitation total, right: annual average temperature (source: own illustration)

3.1.2.4   Results of the Assessment for German Ecosystem Type Classes

The application of the criteria according to Sect. 2 and their quantification in accordance with Sect. 3.1.2 to the ecosystem type classes led to the following allocation of assessment points for the 3 ecosystem services analysed in depth (Table 3.11). The rule-based assessment of all ecosystem services included in CICES according to Haines-Young and Potschin (2018) is shown in Table 3.12.

3.1.2.5   Resulting Maps of the Assessed Ecosystem Services Under Afforestation and Nitrogen Deposition

For the three ecosystem services classified in depth on the basis of rules, an allocation of the integrated assessment according to Table 3.11. Each polygon, which was formed from the intersection of the soil map, climate grid map and Corine Land Cover and assigned to an ecosystem type class, could now be assigned an assessment point for each of the three ecosystem services. The methodological principle of transferring the integrated assessment points of the ecosystem services to the mapped polygons of the 78 German ecosystem type classes is shown in Fig. 3.8.

The spatial distribution of the values assessed for the ecosystem services of the 78 ecosystem type classes in their current state is shown in Figs. 3.9, 3.10 and 3.11 for Germany. For comparison, Figs. 3.12, 3.13 and 3.14 show the assessment results for pristine natural forest vegetation (Table 3.6).

The comparison of the area shares in the assessment levels of the same ecosystem service of pristine natural woody vegetation in the ecosystem type class with the current vegetation of the ecosystem type class (Table 3.6) shows significant current losses of ecosystem services (Table 3.13).

The biggest difference is in the habitat services. This results from the large-scale conversion of potentially natural deciduous and mixed forests into species-poor, unstructured coniferous forest plantations. However, biomass production would also have been higher in natural deciduous or mixed forests than in coniferous forest plantations. The growth-promoting influence of nitrogen deposition has so far led to a drastic increase in primary biomass production in the lowland and low mountain regions of Germany. This also applies to carbon storage. After afforestation in conifer plantations, the proportions of the middle assessment levels in the near-natural deciduous and mixed forests fall to the lower assessment levels.

Table 3.11 Evaluation of the habitat, biomass production and carbon sequestration services of the forest ecosystem type classes

Ecosystem type class_code	Ecosystem type class_name	Hemeroby	Compositional completeness	Habitat value for fauna	Vulnerability/need for protection	Restorability of habitats	Maturity level	Biotope network	Integrative evaluation of the habitat service	Plant physiological production	Water balance of the soil	Nutrient balance	Structure of the floor	Climate-specific productivity	Integrative evaluation the productivity of biomass	Clay content	Mass of litter and crop residues	Yield potential of the soil	pH value in humus	Climate influence	Soil water contamination	Decomposition time	Integrative evaluation the sequestration of carbon
1	Central European to subcontinental, dry, nutrient-poor pine forest	2	2	2	5	3	2	1	2	3	1	1	3	4	1.6	0	1	1	1	4	1	3	0.6
2	Suboceanic, dry, nutrient-poor pine wood	5	5	4	5	3	4	3	5	3	1	1	3	5	1.6	0	1	1	1	5	1	3	0.6
3	Central European to subcontinental, dry, nutrient-poor pine wood	5	5	4	5	3	4	3	5	3	1	1	3	3	1.5	0	1	1	1	3	1	3	0.5
4	Suboceanic, moderately dry to fresh, nutrient-poor pine forest	2	2	2	0	3	2	3	2	3	4	1	3	5	3.1	0	1	1	1	5	4	3	0.8
5	Suboceanic, moderately dry to fresh, poor larch forest	1	1	2	0	4	2	1	1	2	3	1	3	5	2.6	0	2	1	1	5	3	5	0.9
6	Suboceanic, moderately dry to fresh, nutrient-poor beech forest / woodland	5	5	4	5	4	4	3	5	5	3	1	3	5	2.8	1	5	1	1	5	3	3	1.0

(continued)

Table 3.11 (continued)

Ecosystem type class_code	Ecosystem type class_name	Hemeroby	Compositional completeness	Habitat value for fauna	Vulnerability/need for protection	Restorability of habitats	Maturity level	Biotope network	Integrative evaluation of the habitat service	Plant physiological production	Water balance of the soil	Nutrient balance	Structure of the floor	Climate-specific productivity	Integrative evaluation the productivity of biomass	Clay content	Mass of litter and crop residues	Yield potential of the soil	pH value in humus	Climate influence	Soil water contamination	Decomposition time	Integrative evaluation the sequestration of carbon
7	Suboceanic, moderately dry to fresh, nutrient-poor spruce wood	5	3	2	5	3	4	3	3	5	5	1	1	2	3.3	0	3	1	1	2	5	3	0.7
8	Suboceanic, moderately dry to fresh, nutrient-poor spruce forest	1	1	2	0	3	2	1	1	5	4	1	3	5	3.3	0	3	1	1	5	4	3	0.8
9	Suboceanic, moist, nutrient-poor pine forest	2	2	2	1	3	2	3	2	3	4	1	3	5	3.1	1	1	4	1	5	4	3	1.4
10	Suboceanic, moist, fairly nutrient-poor pine forest	3	3	2	0	3	3	3	2	3	4	2	3	5	3.4	2	1	4	2	5	4	3	1.8
11	Central European to subcontinental, moist, nutrient-poor spruce wood	5	4	4	4	3	4	3	4	5	4	1	3	3	3.1	0	3	1	1	3	4	3	0.7
12	Central European to subcontinental, dry, fairly nutrient-poor sessile oak wood	5	5	4	5	4	4	3	5	4	2	2	3	5	2.4	1	5	2	2	5	2	3	1.6

	Description																						
13	Central European to subcontinental, dry, fairly nutrient-poor pine forest	3	3	3	0	3	3	3	**3**	3	2	3	1	3	**2.0**	0	1	1	1	3	2	3	**0.6**
14	Suboceanic, moderately dry to fresh, fairly nutrient-poor beech forest/woodland	5	5	4	5	4	4	3	**5**	4	3	2	3	5	**2.9**	1	1	2	2	5	3	3	**1.2**
15	Moderate boreal climate, moderately dry to fresh, quite nutrient-poor beech wood/forest	3	5	3	5	4	4	3	**4**	3	5	2	1	2	**3.4**	1	4	2	2	2	5	3	**1.5**
16	Suboceanic, moist, fairly nutrient-poor English oak forest	5	5	4	4	4	4	3	**5**	4	4	2	3	5	**3.4**	1	5	2	2	5	4	3	**1.8**
17	Suboceanic, moist, fairly nutrient-poor fir forest	3	5	3	5	3	3	3	**4**	3	3	2	2	4	**2.7**	1	3	2	2	4	3	3	**1.4**
18	Central European to subcontinental, dry, moderately nutrient-rich pine forest	1	1	2	0	3	3	3	**1**	3	2	3	3	3	**2.5**	2	1	3	3	3	2	3	**1.4**
19	Central European to subcontinental, dry, moderately nutritious sessile oak wood	5	5	4	4	4	4	4	**5**	4	2	3	3	4	**2.6**	2	5	3	3	4	2	3	**2.5**
20	Suboceanic, moderately dry to fresh, moderately nutrient-rich pine forest	2	1	2	0	3	3	3	**1**	3	3	3	3	5	**3.1**	2	1	3	3	5	3	3	**1.6**
21	Central European to subcontinental, moderately dry to fresh, moderately nutrient-rich pine forests	1	1	2	0	3	3	3	**1**	3	2	2	3	3	**2.5**	2	1	3	3	3	2	3	**1.4**

(continued)

Table 3.11 (continued)

Ecosystem type class_code	Ecosystem type class_name	Hemeroby	Compositional completeness	Habitat value for fauna	Vulnerability/need for protection	Restorability of habitats	Maturity level	Biotope network	Integrative evaluation of the habitat service	Plant physiological production	Water balance of the soil	Nutrient balance	Structure of the floor	Climate-specific productivity	Integrative evaluation the productivity of biomass	Clay content	Mass of litter and crop residues	Yield potential of the soil	pH value in humus	Climate influence	Soil water contamination	Decomposition time	Integrative evaluation the sequestration of carbon
22	Suboceanic, moist, moderately nutrient-rich oak forest	4	4	3	4	4	4	3	4	4	4	3	3	5	3.7	2	5	3	3	5	4	3	2.7
23	Suboceanic, dry, moderately nutrient-rich, carbonate-rich pine forests of the Alpine valleys	5	5	3	4	3	4	3	4	3	2	3	3	5	2.6	5	3	3	3	5	2	3	2.8
24	Suboceanic, moderately dry to fresh, moderately nutrient-rich beech forest/wood	5	5	3	5	4	4	3	4	5	3	3	3	5	3.3	2	5	3	3	5	3	3	2.6
25	Central European to subcontinental, moderately dry to fresh, moderately nutritious sessile oak-beech wood	5	5	4	5	4	4	3	5	5	3	3	3	4	3.2	2	5	3	3	4	3	3	2.6

No.	Description																							
26	Central European to subcontinental, moderately dry to fresh, moderately nutrient-rich Linden-hornbeam wood	5	5	4	5	4	4	3	**5**	5	3	3	3	3	**3.1**	2	4	3	3	3	3	3	2	**2.2**
27	Moderate boreal climate, moderately dry to fresh, moderately nutrient-rich beech wood/forest stand	5	4	4	5	4	4	3	**4**	5	5	3	1	2	**3.8**	2	5	3	3	3	2	5	3	**2.6**
28	Suboceanic, moderately dry to fresh, moderately nutrient-rich spruce forest	1	1	2	0	3	2	1	**1**	5	4	3	3	5	**3.8**	2	3	3	3	3	5	4	3	**2.2**
29	Suboceanic, moderately dry to fresh, moderately nutritious spruce-fir wood	3	3	2	2	3	3	3	**3**	5	4	3	2	4	**3.6**	2	3	3	3	3	4	4	3	**2.1**
30	Suboceanic, moderately dry to fresh, moderately nutrient-rich Douglas fir forest	1	1	3	0	3	2	3	**2**	3	4	3	2	4	**3.4**	2	2	3	3	3	4	4	4	**1.9**
31	Suboceanic, moderately dry to fresh, moderately nutrient-rich fir-beech forest	5	5	3	5	4	4	3	**4**	5	5	3	2	4	**4.1**	2	5	3	3	3	4	5	3	**2.7**
32	Suboceanic, moderately dry to fresh, moderately nutritious English oak-beech wood	5	5	3	5	4	4	3	**4**	5	4	3	3	5	**3.8**	2	5	3	3	3	5	4	3	**2.7**
33	Suboceanic, moderately dry to fresh, moderately nutrient-rich oak forest	4	4	3	4	4	4	3	**4**	4	4	3	3	5	**3.7**	2	5	3	3	3	5	4	3	**2.7**

(continued)

Table 3.11 (continued)

Ecosystem type class_code	Ecosystem type class_name	Hemeroby	Compositional completeness	Habitat value for fauna	Vulnerability/need for protection	Restorability of habitats	Maturity level	Biotope network	Integrative evaluation of the habitat service	Plant physiological production	Water balance of the soil	Nutrient balance	Structure of the floor	Climate-specific productivity	Integrative evaluation the productivity of biomass	Clay content	Mass of litter and crop residues	Yield potential of the soil	pH value in humus	Climate influence	Soil water contamination	Decomposition time	Integrative evaluation the sequestration of carbon
34	Moderate boreal climate, moist, moderately nutrient-rich fir-beech forest	5	5	3	5	4	4	3	**4**	5	4	3	1	2	**3.3**	2	3	3	3	2	4	3	**2.0**
35	Suboceanic, moist, moderately nutrient-rich fir-beech forest	5	5	3	5	4	4	3	**4**	5	4	3	2	4	**3.6**	2	5	3	3	4	4	3	**2.6**
36	Suboceanic, moist, moderately nutrient-rich oak-hornbeam forest	5	5	4	5	4	4	3	**5**	5	4	3	3	5	**3.8**	2	5	3	3	5	4	3	**2.7**
37	Suboceanic, moist, moderately nutritious fir wood	3	3	2	4	3	3	3	**3**	3	3	3	2	4	**2.9**	2	3	3	3	4	3	3	**2.1**
38	Suboceanic, dry, moderately nutrient-rich, carbonate pine forest	5	3	4	5	3	3	3	**4**	3	2	3	3	5	**2.6**	5	1	3	3	5	2	3	**2.3**

No.	Description																						
39	Suboceanic, moderately dry to fresh, moderately nutrient-rich carbonate spruce forest	2	1	2	0	3	2	1	**1**	5	4	3	3	5	**3.8**	5	5	3	3	5	4	3	**2.9**
40	Central European to subcontinental, dry, vigorous, nutrient-rich rock maple - sessile oak wood	5	5	4	5	4	4	3	**5**	4	1	4	3	3	**2.3**	3	5	4	4	3	1	3	**3.3**
41	Central European to subcontinental, moderately dry to fresh, vigorous, nutrient-rich hornbeam forest	5	5	4	5	4	4	3	**5**	5	3	4	3	4	**3.4**	3	5	4	4	4	3	3	**3.5**
42	Central European to subcontinental, moderately dry to fresh, vigorous, nutrient-rich winter lime wood	5	5	4	5	4	4	3	**5**	5	4	4	3	4	**3.9**	3	5	4	4	4	4	3	**3.6**
43	Moderate boreal climate, moderately dry to fresh, strong, nutritious spruce-fir wood	5	3	3	5	4	4	3	**4**	5	5	4	1	2	**4.1**	3	3	4	4	2	5	3	**2.8**
44	Suboceanic, moderately dry to fresh, nutrient-rich beech forest	5	5	3	5	4	4	3	**4**	5	5	4	2	4	**4.3**	3	5	4	4	4	5	3	**3.6**
45	Suboceanic, moist, fast-growing, nutrient-rich sycamore and ash forest of the montane stage	5	5	4	5	3	4	3	**5**	5	4	4	3	5	**4.0**	3	5	4	4	5	4	1	**3.5**

(continued)

Table 3.11 (continued)

Ecosystem type class_code	Ecosystem type class_name	Hemeroby	Compositional completeness	Habitat value for fauna	Vulnerability/need for protection	Restorability of habitats	Maturity level	Biotope network	Integrative evaluation of the habitat service	Plant physiological production	Water balance of the soil	Nutrient balance	Structure of the floor	Climate-specific productivity	Integrative evaluation the productivity of biomass	Clay content	Mass of litter and crop residues	Yield potential of the soil	pH value in humus	Climate influence	Soil water contamination	Decomposition time	Integrative evaluation the sequestration of carbon
46	Suboceanic, moderately dry to fresh, vigorous, nutrient-rich oak forest	3	4	3	4	4	3	3	**4**	4	4	4	3	5	**3.9**	3	5	4	4	5	4	3	**3.6**
47	Suboceanic, moist, vigorous, nutritious beech forest	5	4	3	5	4	4	3	**4**	5	4	4	3	5	**4.0**	3	5	4	4	5	4	3	**3.6**
48	Suboceanic, moist, nutrient-rich oak-hornbeam-ash wood	5	5	4	5	4	4	3	**5**	5	4	4	3	5	**4.0**	3	5	4	4	5	4	3	**3.6**
49	Suboceanic, moist, nutrient-rich black alder floodplain forest	5	5	4	5	4	4	5	**5**	4	2	4	4	4	**3.0**	3	5	4	4	4	5	1	**3.5**
50	Suboceanic, moist, flooded, vigorous, nutrient-rich elm and English oak floodplain forest	5	5	4	5	4	4	5	**5**	4	4	4	4	5	**4.1**	3	5	4	4	5	4	3	**3.6**

No.	Description																					
51	Suboceanic moderately dry to fresh, nutrient-rich oak with alternating dry wood	5	4	4	4	4	**5**	3	5	4	3	5	3	5	**3.7**	4	5	5	5	5	3	3	**4.5**
52	Central European to subcontinental, moderately dry to fresh, nutrient-rich winter lime wood	5	4	5	4	4	**5**	3	5	5	3	5	3	3	**3.6**	4	4	5	5	3	3	2	**3.8**
53	Temperate boreal climate, moderately dry to fresh, nutrient-rich sycamore beech forest	5	3	5	4	4	**4**	3	4	5	5	5	1	2	**4.3**	4	5	5	5	2	5	3	**4.4**
54	Suboceanic, moderately dry to fresh, nutrient-rich hornbeam–beech forest	5	3	5	4	4	**4**	3	4	5	4	5	2	4	**4.1**	4	5	5	5	4	4	3	**4.5**
55	Central European to subcontinental, moderately dry to fresh, nutrient-rich hornbeam forests	5	3	5	4	3	**4**	3	4	5	4	5	3	4	**4.2**	4	5	5	5	4	4	3	**4.5**
56	Suboceanic, moderately dry to fresh, nutrient-rich beech forest	5	4	5	4	4	**4**	3	4	5	5	5	2	4	**4.6**	4	4	5	5	4	5	3	**4.6**
57	Suboceanic, moderately dry to fresh, nutrient-rich mountain elm-summer linden-blockwood trees	5	4	5	4	4	**5**	3	5	5	4	5	3	5	**4.3**	4	4	5	5	5	4	1	**3.9**
58	Suboceanic, moist, nutrient-rich beech forest	5	3	5	4	4	**4**	3	4	5	4	5	2	4	**4.1**	4	5	5	5	4	4	3	**4.5**

(continued)

Table 3.11 (continued)

Ecosystem type class_code	Ecosystem type class_name	Hemeroby	Compositional completeness	Habitat value for fauna	Vulnerability/need for protection	Restorability of habitats	Maturity level	Biotope network	Integrative evaluation of the habitat service	Plant physiological production	Water balance of the soil	Nutrient balance	Structure of the floor	Climate-specific productivity	Integrative evaluation the productivity of biomass	Clay content	Mass of litter and crop residues	Yield potential of the soil	pH value in humus	Climate influence	Soil water contamination	Decomposition time	Integrative evaluation the sequestration of carbon
59	Suboceanic, moist, nutrient-rich ash wood	5	5	4	5	4	4	3	5	5	4	5	3	5	4.3	4	4	5	5	5	4	2	4.0
60	Suboceanic, moist, flooded, nutrient-rich floodplain forest with Salix x rubens	5	5	4	5	3	4	5	5	2	3	5	4	5	3.7	4	4	5	4	5	4	2	3.9
61	Suboceanic, dry, nutrient-rich, carbonate-rich sessile oak dry forest	5	5	4	5	4	4	3	5	4	1	5	3	5	2.7	5	5	5	2	5	1	3	4.4
62	Suboceanic, dry, nutrient-rich, carbonate-containing oak wood	5	5	4	5	4	4	3	5	4	2	5	3	5	3.2	5	5	5	5	5	2	3	4.7
63	Suboceanic, moderately dry to fresh, nutrient-rich, carbonate-rich sunny slope beech forest	5	5	4	5	4	4	3	5	5	3	5	3	5	3.8	5	5	5	5	5	3	3	4.8

64	Suboceanic, moderately dry to fresh, nutrient-rich, carbonate-rich beech wood/forest stand	5	4	5	4	4	3	**5**	5	5	4	5	3	5	**4.3**	5	5	5	5	5	4	3	**4.8**
65	Suboceanic, moist, nutrient-rich, carbonate-containing ash wood	5	4	5	5	4	3	**5**	4	5	4	5	3	5	**4.2**	5	5	5	5	5	4	1	**4.7**
66	Moderate boreal climate, very wet, nutrient-poor organic raised bog	5	4	5	5	4	3	**5**	0	2	1	2	1	2	**1.2**	2	5	2	2	2	5	5	**2.0**
67	Suboceanic, very wet, nutrient-poor organic raised bog	5	4	5	5	4	3	**5**	0	1	1	1	3	5	**1.4**	1	2	1	1	5	5	5	**1.3**
68	Moderate boreal climate, moist, fairly nutrient-poor organic raised bog wood	5	4	5	3	4	3	**5**	1	3	3	3	1	1	**2.5**	3	1	3	3	1	3	3	**1.6**
69	Moderate boreal climate, humid, fairly nutrient-poor organic Carpathian birch wood	5	4	5	3	4	3	**5**	2	3	3	3	1	2	**2.6**	3	4	3	3	2	4	2	**2.4**
70	Central European to subcontinental, moist, fairly nutrient-poor organic bog birch wood	5	4	5	3	4	3	**5**	2	3	2	3	3	5	**2.6**	3	3	3	3	5	5	5	**2.6**
71	Suboceanic, moist, moderately nutritious organic black alder wood	5	4	5	4	4	3	**5**	4	4	4	4	3	5	**3.9**	4	5	4	4	5	4	1	**3.8**

(continued)

Table 3.11 (continued)

Ecosystem type class_code	Ecosystem type class_name	Hemeroby	Compositional completeness	Habitat value for fauna	Vulnerability/need for protection	Restorability of habitats	Maturity level	Biotope network	Integrative evaluation of the habitat service	Plant physiological production	Water balance of the soil	Nutrient balance	Structure of the floor	Climate-specific productivity	Integrative evaluation the productivity of biomass	Clay content	Mass of litter and crop residues	Yield potential of the soil	pH value in humus	Climate influence	Soil water contamination	Decomposition time	Integrative evaluation the sequestration of carbon
72	Suboceanic, moist, moderately nutritious organic black alder wood	5	5	4	5	3	4	3	**5**	4	2	4	3	5	**2.9**	4	5	4	4	5	5	1	**3.8**
73	Suboceanic, moist, strong, nutritious organic black alder wood	5	5	4	5	3	4	3	**5**	4	2	5	3	5	**3.2**	5	5	5	5	5	5	1	**4.8**
74	Suboceanic, moist flooded, vigorous, nutrient-rich organic grey alder forest	5	5	4	5	3	4	5	**5**	4	3	5	4	4	**3.8**	5	4	5	5	4	4	2	**4.2**
75	Suboceanic, moist flooded, vigorous, nutritious organic Prunus padus ash wood	5	5	4	5	4	4	3	**5**	4	3	5	3	5	**3.7**	5	5	5	5	5	4	1	**4.7**

76	Temperate boreal climate, moderately dry to fresh, fairly nutrient-poor carbonate-containing mountain pine krummholz	5	5	3	5	4	4	3	**4**	1	5	2	1	1	**3.3**	2	3	2	2	1	5	3	**1.6**
77	Temperate boreal climate, moderately dry to fresh, fairly nutrient-poor, carbonate-rich spruce wood	5	5	3	5	3	4	3	**4**	5	5	2	1	2	**3.6**	2	3	2	2	2	5	3	**1.6**
78	Temperate boreal climate, moderately dry to fresh, fairly nutrient-poor carbonate-rich spruce-beech wood	5	5	3	5	4	4	3	**4**	5	5	3	1	2	**3.8**	3	3	3	2	2	5	3	**2.3**

Table 3.12 Assessment of all potential ecosystem services

Ecosystem classes of forests/ woodlands	Regulation and maintenance services						Supply services	Cultural achievements		
	Erosion control function - stabilisation of solids (soil, sand, snow, etc.)	Flood protection function	Function of groundwater recharge/ drinking water supply	Groundwater protection function / safeguarding drinking water quality	Habitat Service - Self-regulation and self-organisation of ecosystems	Service for carbon coating	Biomass primary productivity - plant-based raw materials	Nature experience and recreational function	Aesthetic perception, nature education, natural heritage	Preservation function of cultural heritage, legacy for future generations
1 Central European to subcontinental, dry, nutrient-poor pine forest	3.5	0	3	3	2.5	1	2	1	0.5	0
2 Suboceanic, dry, nutrient-poor pine forest	3	1	3	3	5	1	2	5	4	4
3 Central European to subcontinental, dry, nutrient-poor pine wood	3.5	0	3	3	4.5	1	2	3	2.5	3.5
4 Suboceanic, moderately dry to fresh, nutrient-poor pine forest	4	0	3	3	2	1	3	1	0	0

5	Suboceanic, moderately dry to fresh, poor larch forest	2	0	3	2	1	1	3	1	1	0	0
6	Suboceanic, moderately dry to fresh, nutrient-poor beech forest / woodland	1	0	4	1	5	1	3	5	5	5	4
7	Suboceanic, moderately dry to fresh, nutrient-poor spruce wood	1	0	3	3	3	1	3	3	3	2	1
8	Suboceanic, moderately dry to fresh, nutrient-poor spruce forest	2	0	3	3	1	1	3	1	1	0	0
9	Suboceanic, moist, nutrient-poor pine forest	5	0	3	2	2	1	3	0	0	0	0

(continued)

Table 3.12 (continued)

Ecosystem classes of forests/ woodlands	Regulation and maintenance services						Supply services	Cultural achievements		
	Erosion control function - stabilisation of solids (soil, sand, snow, etc.)	Flood protection function	Function of groundwater recharge/ drinking water supply	Groundwater protection function / safeguarding drinking water quality	Habitat Service - Self-regulation and self-organisation of ecosystems	Service for carbon coating	Biomass primary productivity - plant-based raw materials	Nature experience and recreational function	Aesthetic perception, nature education, natural heritage	Preservation function of cultural heritage, legacy for future generations
10 Suboceanic, moist, fairly nutrient-poor pine forest	5	0	3	2	2	1.5	3	0	0	0
11 Central European to subcontinental, moist, nutrient-poor spruce wood	4	0	3	2	3.5	1.5	3	0	1	0.5
12 Central European to subcontinental, dry, fairly nutrient-poor sessile oak wood	4	0	2.5	2.5	3	1.5	2	5	4	3

13	Central European to subcontinental, dry, fairly nutrient-poor pine forest	4	0	3	3	4	1	2	1	0	1
14	Suboceanic, moderately dry to fresh, fairly nutrient-poor beech forest/woodland	4	0	4	1	5	1.5	3	5	5	4
15	Moderate boreal climate, moderately dry to fresh, quite nutrient-poor beech wood/forest	2	0	4	1	4	1.3	3.75	5	3	3
16	Suboceanic, moist, fairly nutrient-poor English oak forest	5	0	4	1	5	2	3	2	5	4
17	Suboceanic, moist, fairly nutrient-poor fir forest	4	0	3	2	4	1	3	2	2	4

(continued)

Table 3.12 (continued)

Ecosystem classes of forests/ woodlands	Regulation and maintenance services						Supply services	Cultural achievements		
	Erosion control function - stabilisation of solids (soil, sand, snow, etc.)	Flood protection function	Function of groundwater recharge/ drinking water supply	Groundwater protection function / safeguarding drinking water quality	Habitat Service - Self- regulation and self- organisation of ecosystems	Service for carbon coating	Biomass primary productivity - plant-based raw materials	Nature experience and recreational function	Aesthetic perception, nature education, natural heritage	Preservation function of cultural heritage, legacy for future generations
18 Central European to subcontinental, dry, moderately nutrient-rich pine forest	4	0	3	4	1	1.5	3	1	0	0
19 Central European to subcontinental, dry, moderately nutritious sessile oak wood	4.5	0	2.5	2.5	5	2	3	5	4	3
20 Suboceanic, moderately dry to fresh, moderately nutrient-rich pine forest	4	0	3	4	1	2	3.6	1	0	0

#	Description										
21	Central European to subcontinental, moderately dry to fresh, moderately nutrient-rich pine forests	4	0	3	4	1.5	1.5	3	1	0	0.5
22	Suboceanic, moist, moderately nutrient-rich oak forest	5	3	4	1	4	3	4	2	4	2
23	Suboceanic, dry, moderately nutrient-rich, carbonate-rich pine forests of the Alpine valleys	4	0	3	4	4	3	3	3	3	2
24	Suboceanic, moderately dry to fresh, moderately nutrient-rich beech forest/wood	4	0	4	1	1.75	3	3.5	4	4	2.5

(continued)

Table 3.12 (continued)

| Ecosystem classes of forests/ woodlands | Regulation and maintenance services | | | | | | Supply services | Cultural achievements | | |
	Erosion control function - stabilisation of solids (soil, sand, snow, etc.)	Flood protection function	Function of groundwater recharge/ drinking water supply	Groundwater protection function / safeguarding drinking water quality	Habitat Service - Self-regulation and self-organisation of ecosystems	Service for carbon coating	Biomass primary productivity - plant-based raw materials	Nature experience and recreational function	Aesthetic perception, nature education, natural heritage	Preservation function of cultural heritage, legacy for future generations
25 Central European to subcontinental, moderately dry to fresh, moderately nutritious sessile oak-beech wood	5	0	3	2	5	3	3	5	5	3
26 Central European to subcontinental, moderately dry to fresh, moderately nutrient-rich lime-hornbeam wood	3	0	2	3	5	2	3	5	5	5

No.	Description										
27	Moderate boreal climate, moderately dry to fresh, moderately nutrient-rich beech wood/forest stand	3	0	4	1	4	2.5	4	4.5	2.5	2
28	Suboceanic, moderately dry to fresh, moderately nutrient-rich spruce forest	2	0	3	2	2	2	4	1	0	0
29	Suboceanic, moderately dry to fresh, moderately nutritious spruce-fir wood	3	0	3	2	3	2	4	3	2	1
30	Suboceanic, moderately dry to fresh, moderately nutrient-rich Douglas fir forest	2	0	3	2	2	2	3	1	0	0

(continued)

Table 3.12 (continued)

Ecosystem classes of forests/ woodlands	Regulation and maintenance services						Supply services	Cultural achievements		
	Erosion control function - stabilisation of solids (soil, sand, snow, etc.)	Flood protection function	Function of groundwater recharge/ drinking water supply	Groundwater protection function / safeguarding drinking water quality	Habitat Service - Self-regulation and self-organisation of ecosystems	Service for carbon coating	Biomass primary productivity - plant-based raw materials	Nature experience and recreational function	Aesthetic perception, nature education, natural heritage	Preservation function of cultural heritage, legacy for future generations
31 Suboceanic, moderately dry to fresh, moderately nutrient-rich fir-beech forest	3.5	0	4	1	4	3	4	4.5	4.5	2.5
32 Suboceanic, moderately dry to fresh, moderately nutritious English oak-beech wood	4	0	4	1	4	3	4	5	5	3
33 Suboceanic, moderately dry to fresh, moderately nutrient-rich oak forest	5	0	4	1	4	3	4	3	2	0

34	Moderate boreal climate, moist, moderately nutrient-rich fir-beech forest	4	0	4	1	4	2	3	3	2	3	4	2
35	Suboceanic, moist, moderately nutrient-rich fir-beech forest	4	0	4	1	4	3	4	2	3	2	5	3
36	Suboceanic, moist, moderately nutrient-rich oak-hornbeam forest	4	3	4	1	5	3	4	3	3	3	5	3
37	Suboceanic, moist, moderately nutritious fir wood	4	0	3	2	3	2	3	2	2	2	2	4
38	Suboceanic, dry, moderately nutrient-rich, carbonate pine forest	4	0	3	4	4	2	3	2	2	2	2	1

(continued)

Table 3.12 (continued)

Ecosystem classes of forests/ woodlands	Regulation and maintenance services						Supply services	Cultural achievements		
	Erosion control function - stabilisation of solids (soil, sand, snow, etc.)	Flood protection function	Function of groundwater recharge/ drinking water supply	Groundwater protection function / safeguarding drinking water quality	Habitat Service - Self-regulation and self-organisation of ecosystems	Service for carbon coating	Biomass primary productivity - plant-based raw materials	Nature experience and recreational function	Aesthetic perception, nature education, natural heritage	Preservation function of cultural heritage, legacy for future generations
39 Suboceanic, moderately dry to fresh, moderately nutrient-rich carbonate spruce forest	3	0	3	2	1	3	4	1	0	0
40 Central European to subcontinental, dry, vigorous, nutritious rock maple-cone oak wood	2	0	1	4	5	3	2	4	5	4

41	3	Central European to subcontinental, moderately dry to fresh, vigorous, nutrient-rich hornbeam forest	0	3	2	5	4	3	5	5	4
42	3	Central European to subcontinental, moderately dry to fresh, vigorous, nutrient-rich winter lime wood	0	2.5	2.5	5	3.5	3.5	5	5	4.5
43	4	Moderate boreal climate, moderately dry to fresh, strong, nutritious spruce-fir wood	0	4	1	4	3	4	5	3	2
44	3	Suboceanic, moderately dry to fresh, nutrient-rich beech forest	0	4	1	4	4	4	4	2	2

(continued)

Table 3.12 (continued)

Ecosystem classes of forests/ woodlands	Regulation and maintenance services						Supply services	Cultural achievements		
	Erosion control function - stabilisation of solids (soil, sand, snow, etc.)	Flood protection function	Function of groundwater recharge/ drinking water supply	Groundwater protection function / safeguarding drinking water quality	Habitat Service - Self-regulation and self-organisation of ecosystems	Service for carbon coating	Biomass primary productivity - plant-based raw materials	Nature experience and recreational function	Aesthetic perception, nature education, natural heritage	Preservation function of cultural heritage, legacy for future generations
45 Suboceanic, moist, fast-growing, nutrient-rich sycamore and ash forest of the montane stage	5	5	4	1	5	4	4	2	5	4
46 Suboceanic, moderately dry to fresh, vigorous, nutrient-rich oak forest	4	0	4	1	4	4	4	3	2	0
47 Suboceanic, moist, vigorous, nutritious beech forest	4	0	4	1	4	4	4	2	2	2

48	Suboceanic, moist, nutrient-rich oak-hornbeam-ash wood	5	4	4	1	5	3.5	4	3	5	4.5
49	Suboceanic, moist, nutrient-rich black alder floodplain forest	5	5	4	1	5	3	3	2	5	4
50	Suboceanic, moist, flooded, vigorous, nutrient-rich elm and English oak floodplain forest	5	5	4	1	5	4	4	4	5	4
51	Suboceanic moderately dry to fresh, nutrient-rich oak with alternating dry wood	4	0	4	1	5	5	4	5	4	4

(continued)

Table 3.12 (continued)

| Ecosystem classes of forests/woodlands | Regulation and maintenance services | | | | | | Supply services | Cultural achievements | | |
	Erosion control function - stabilisation of solids (soil, sand, snow, etc.)	Flood protection function	Function of groundwater recharge/drinking water supply	Groundwater protection function / safeguarding drinking water quality	Habitat Service - Self-regulation and self-organisation of ecosystems	Service for carbon coating	Biomass primary productivity - plant-based raw materials	Nature experience and recreational function	Aesthetic perception, nature education, natural heritage	Preservation function of cultural heritage, legacy for future generations
52 Central European to subcontinental, moderately dry to fresh, nutrient-rich winter lime wood	2.5	0	1.5	3.5	5	4	4	5	5	4.5
53 Temperate boreal climate, moderately dry to fresh, nutrient-rich sycamore beech forest	4	0	4	1	4	4	4	5	5	2

54	Suboceanic, moderately dry to fresh, nutrient-rich hornbeam-beech forest	3	0	3	2	4	5	4	5	5	4
55	Central European to subcontinental, moderately dry to fresh, nutrient-rich hornbeam forests	3	0	3	2	4	5	4	5	5	4
56	Suboceanic, moderately dry to fresh, nutrient-rich beech forest	4	0	4	1	4	5	5	4.5	3.5	3.5
57	Suboceanic, moderately dry to fresh, nutrient-rich mountain elm-summer linden-blockwood trees	2	0	3	2	5	4	4	4	5	4

(continued)

Table 3.12 (continued)

Ecosystem classes of forests/ woodlands	Regulation and maintenance services						Supply services	Cultural achievements		
	Erosion control function - stabilisation of solids (soil, sand, snow, etc.)	Flood protection function	Function of groundwater recharge/ drinking water supply	Groundwater protection function / safeguarding drinking water quality	Habitat Service - Self-regulation and self-organisation of ecosystems	Service for carbon coating	Biomass primary productivity - plant-based raw materials	Nature experience and recreational function	Aesthetic perception, nature education, natural heritage	Preservation function of cultural heritage, legacy for future generations
58 Suboceanic, moist, nutrient-rich beech forest	4	0	4	1	4	5	4	2	2.5	3.5
59 Suboceanic, moist, nutrient-rich ash wood	5	5	4	1	5	4	4	3	5	4.5
60 Suboceanic, moist, flooded, nutrient-rich floodplain forest with Salix x rubens	5	5	4	1	5	4	4	2	5	5
61 Suboceanic, dry, nutrient-rich, carbonate-rich sessile oak dry forest	2	0	2	3	5	4	3	4	5	4

62	Suboceanic, dry, nutrient-rich, carbonate-containing oak wood	4	0	4	1	5	5	3	5	4	4
63	Suboceanic, moderately dry to fresh, nutrient-rich, carbonate-rich sunny slope beech forest	2	0	4	1	5	5	4	5	4	4
64	Suboceanic, moderately dry to fresh, nutrient-rich, carbonate-rich beech wood/ forest stand	3.5	0	4	1	5	5	4	4	4	3
65	Suboceanic, moist, nutrient-rich, carbonate-containing ash wood	5	4	4	1	5	5	4	2	5	4
66	Moderate boreal climate, very wet, nutrient-poor organic raised bog	5	0	0	0	5	2	1	0	5	4

(continued)

Table 3.12 (continued)

Ecosystem classes of forests/ woodlands	Regulation and maintenance services					Supply services		Cultural achievements		
	Erosion control function - stabilisation of solids (soil, sand, snow, etc.)	Flood protection function	Function of groundwater recharge/ drinking water supply	Groundwater protection function / safeguarding drinking water quality	Habitat Service - Self-regulation and self-organisation of ecosystems	Service for carbon coating	Biomass primary productivity - plant-based raw materials	Nature experience and recreational function	Aesthetic perception, nature education, natural heritage	Preservation function of cultural heritage, legacy for future generations
67 Suboceanic, very moist, nutrient-poor organic raised bog	5	0	0	0	5	1	1	0	5	4
68 Moderate boreal climate, moist, fairly nutrient-poor organic raised bog wood	5	0	2	0	5	2.5	2.5	0	2.5	4
69 Moderate boreal climate, humid, fairly nutrient-poor organic Carpathian birch forest	5	0	4	1	5	2	3	3	5	5

70	Central European to subcontinental, moist, fairly nutrient-poor organic bog birch wood	5	0	4	1	5	2.75	3	3	5	4
71	Suboceanic, moist, moderately nutritious organic black alder wood	5	0	4	1	5	4	4	2	5	4
72	Suboceanic, moist, moderately nutritious organic black alder wood	5	0	4	1	5	4	3	2	5	4
73	Suboceanic, moist, strong, nutritious organic black alder wood	5	0	4	1	5	5	3.3	2	5	4
74	Suboceanic, moist flooded, vigorous, nutrient-rich organic grey alder forest	5	0	4	1	5	4	4	2	5	4

(continued)

Table 3.12 (continued)

Ecosystem classes of forests/woodlands	Regulation and maintenance services						Supply services	Cultural achievements		
	Erosion control function - stabilisation of solids (soil, sand, snow, etc.)	Flood protection function	Function of groundwater recharge/drinking water supply	Groundwater protection function / safeguarding drinking water quality	Habitat Service - Self-regulation and self-organisation of ecosystems	Service for carbon coating	Biomass primary productivity - plant-based raw materials	Nature experience and recreational function	Aesthetic perception, nature education, natural heritage	Preservation function of cultural heritage, legacy for future generations
75 Suboceanic, moist flooded, vigorous, nutritious organic Prunus padus ash wood	5	0	4	1	5	5	4	2	5	4
76 Moderate boreal climate, moderately dry to fresh, quite nutrient-poor carbonate-containing mountain pine krummholz	1	0	4	1	4	2	3	2	4	4

No.	Description										
77	Temperate boreal climate, moderately dry to fresh, fairly nutrient-poor, carbonate-rich spruce wood	1	0	3	2	4	2	4	3	2	1
78	Temperate boreal climate, moderately dry to fresh, fairly nutrient-poor carbonate-rich spruce-beech wood	2	0	4	1	4	2	4	5	3	2

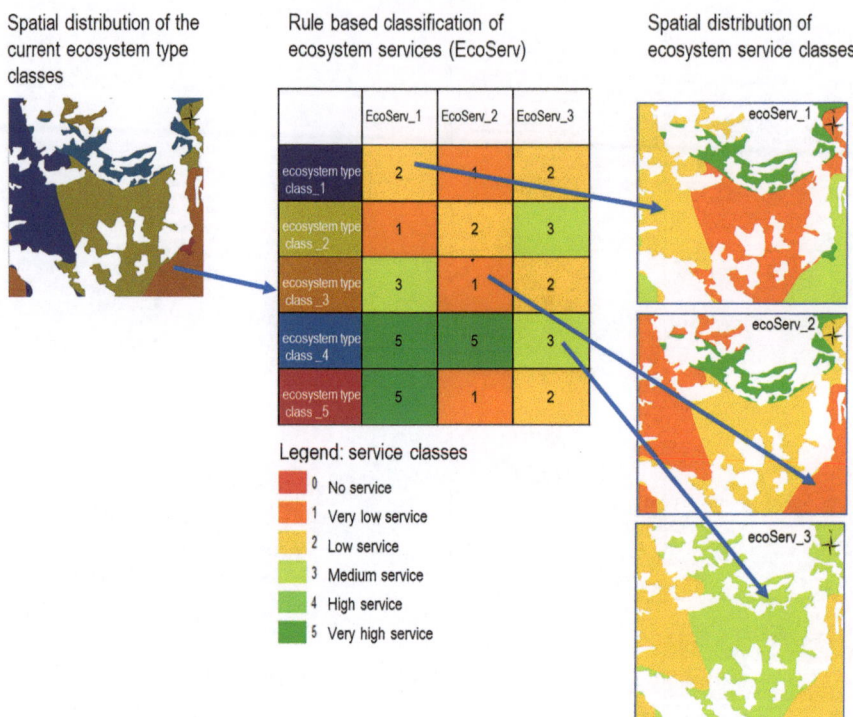

Fig. 3.8 Principle illustration for the mapping of rule-based assessed ecosystem services ("EcoServ") based on the mapped spatial distribution of ecosystem type classes in Germany (source: authors' own illustration)

3.2   Mapping of Special Areas

3.2.1   Modelling Approach and Data Basis for the Kellerwald National Park

The Kellerwald National Park is part of the Kellerwald-Edersee Nature Park, which is located in the north of the federal state of Hesse. It is a UNESCO World Heritage Site for primeval beech forests and covers around 77 km² of the low mountain range in western Hesse. The aim of the Kellerwald-Edersee National Park is to permanently protect the beech forest, which is unique in Western Europe in terms of its size and naturalness. According to the motto "Let nature be nature", the wilderness of tomorrow is to be created here. The national park already fulfils the criteria of the IUCN (International Union for Conservation of Nature), according to which at least 75% of the area should be left to natural dynamics without human intervention (process protection).

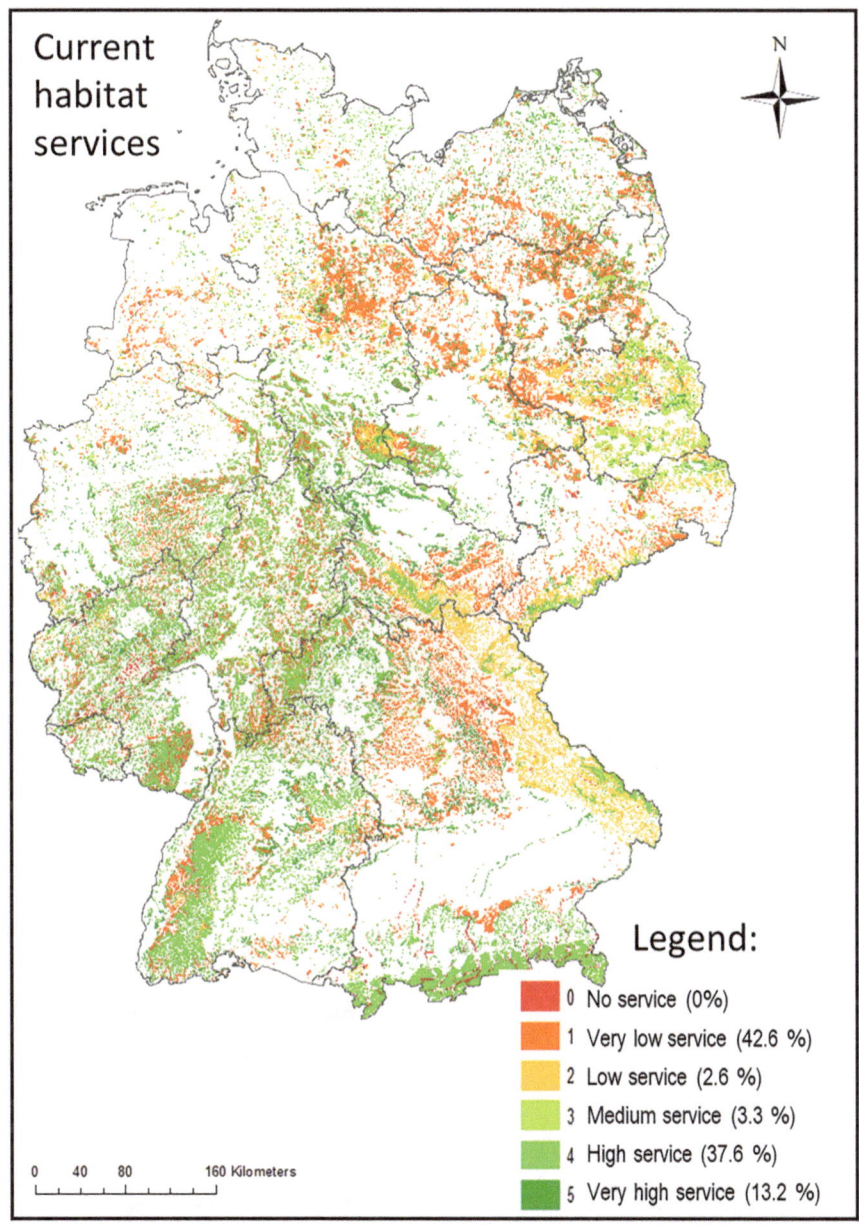

Fig. 3.9 Current habitat service of the ecosystem type classes in Germany. (Source: Authors' own illustration)

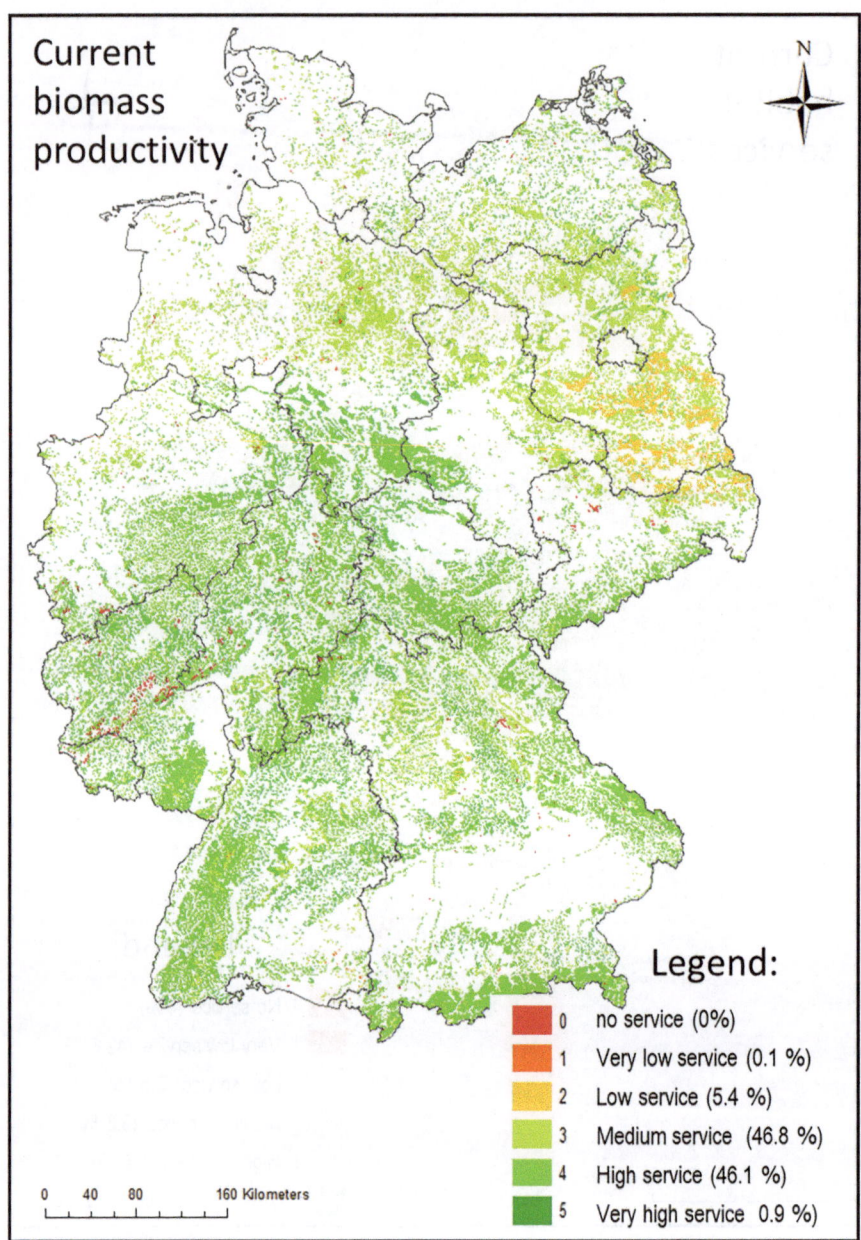

Fig. 3.10 Current biomass primary productivity service of the ecosystem type classes (wood only). (Source: Authors' own illustration)

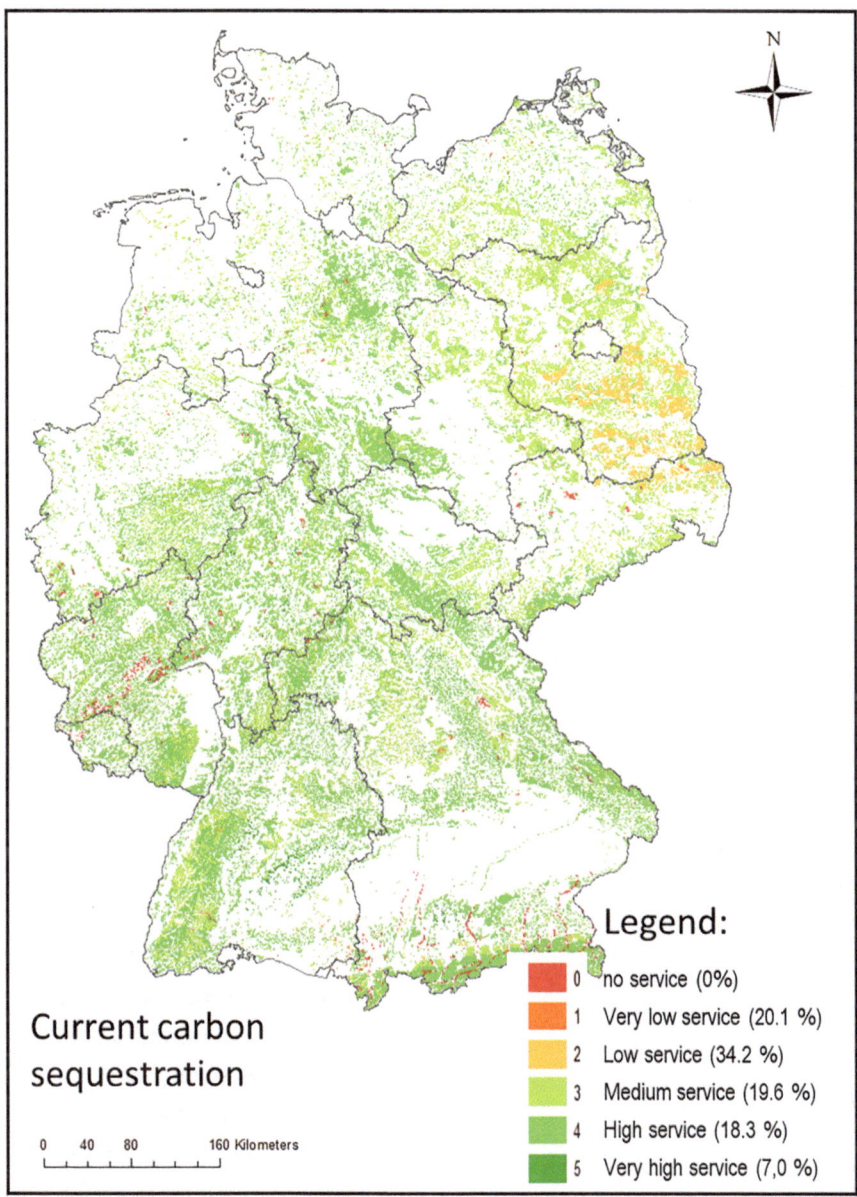

Fig. 3.11 Current carbon sequestration capacity of the ecosystem type classes. (Source: Authors' own illustration)

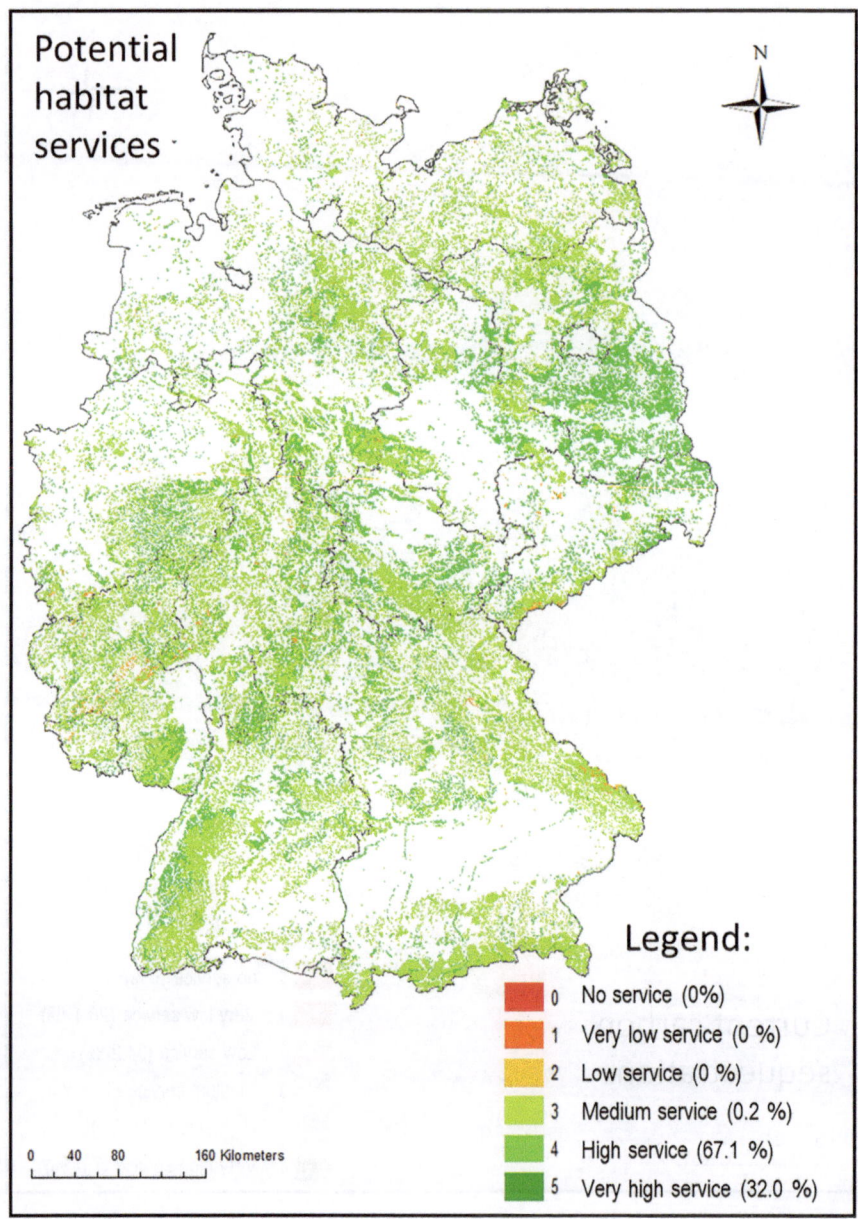

Fig. 3.12 Potential habitat service of pristine natural wood vegetation in the ecosystem type classes in Germany. (Source: Authors' own illustration)

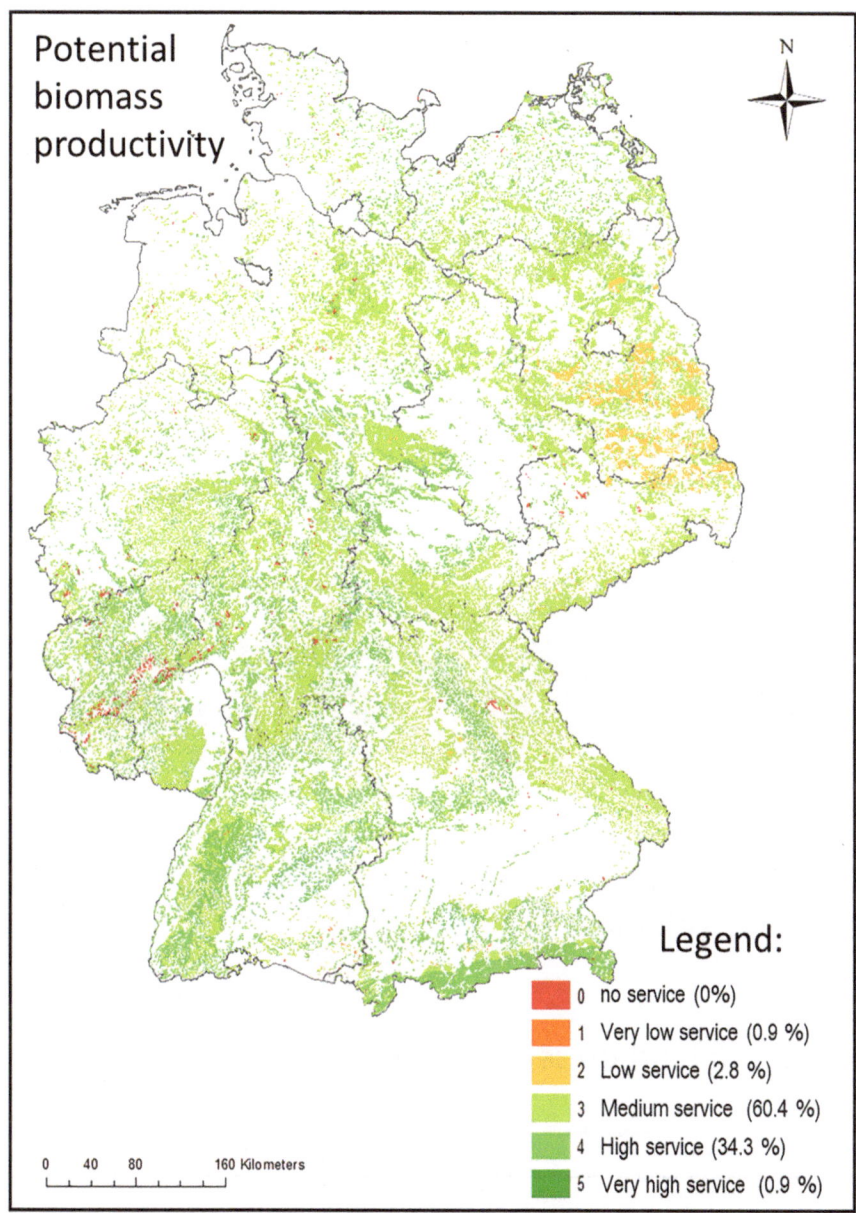

Fig. 3.13 Potential primary biomass productivity of pristine wood vegetation in the ecosystem type classes (timber only). (Source: Author's own illustration)

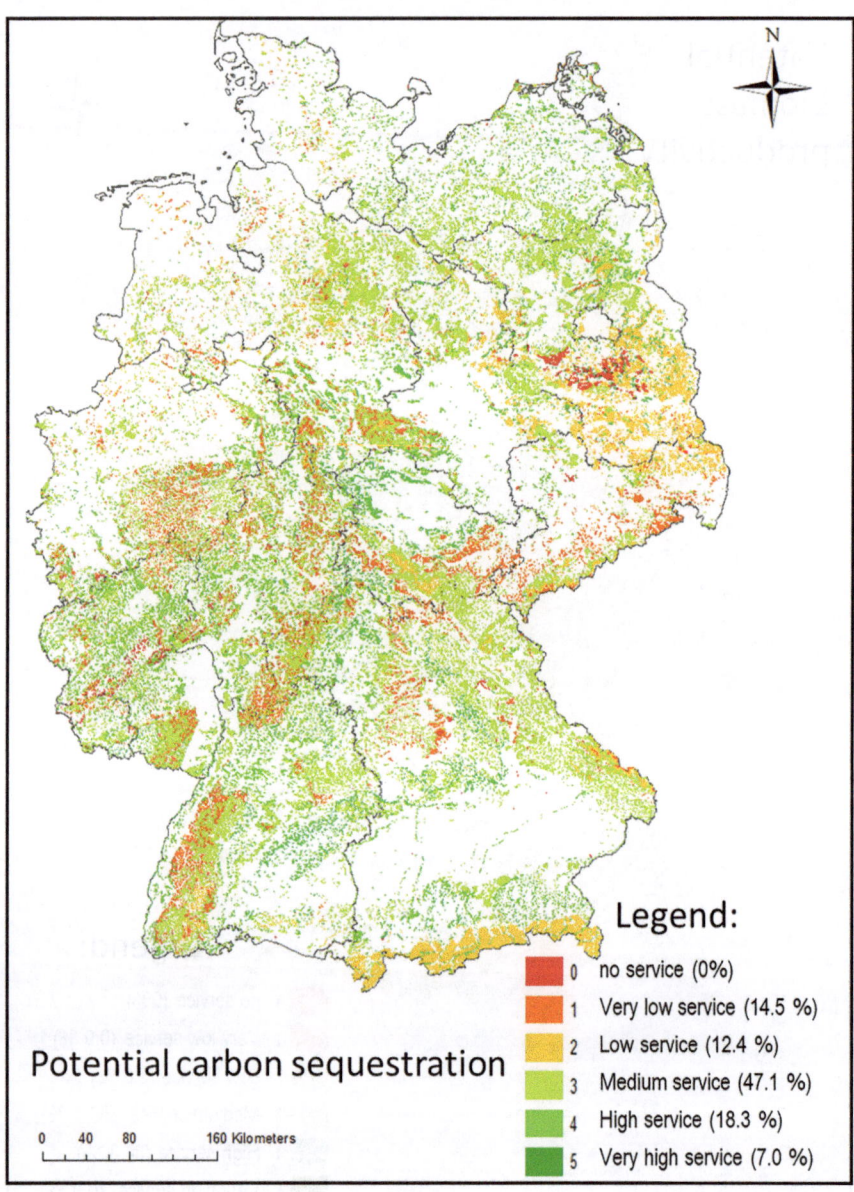

Fig. 3.14 Potential carbon sequestration capacity of pristine natural wood vegetation in the ecosystem type classes. (Source: Authors' own illustration)

Table 3.13 Comparison of the area shares in the assessment levels of the same ecosystem service of the potentially natural ecosystem type class with the current ecosystem type class and the current biomass productivity under the influence of N deposition

Evaluation score	Current habitat service Figure 3.9	Potential habitat service Figure 3.12	Current biomass productivity Figure 3.10	Potential productivity of biomass Figure 3.13	Current carbon sequestration Figure 3.11	Potential carbon storage Figure 3.14
No service	0%	0%	0%	0%	0%	0%
Very little service	42.6%	0%	0.06%	0.9%	20.1%	14.5%
Low service	2.6%	0%	1.4%	2.8%	34.2%	12.4%
Intermediate service	3.3%	0.2%	7.4%	60.4%	19.6%	47.1%
High level of service	37.6%	67.1%	38.8%	34.3%	18.3%	18.3%
Very good service	13.2%	32.0%	52.3%	0.9%	7.0%	7.0%

Orographically, the area is between 194 m and 626 m above sea level and can be described as very hilly. The average annual temperature in the period 1991–2020 was 8.6 °C (DWD 2021). The amount of precipitation in the area varies between 771 and 791 mm/a (DWD 2021). The DeMartonne index is 16.4 and the length of the growing season averages 153 days per year with an average daily temperature of 10 °C or more.

A total of 105 vegetation surveys from the Kellerwald National Park were available as study material (Schröder et al. 2020). 63.4% of the forest area of the national park corresponds to the natural forest vegetation in terms of its tree species composition. These are the various beech forests, the limestone-maple ravine forests and the woodrush-alder forests. A further 7.6% of the forest area comes close to natural forest vegetation. These are near-natural mixed deciduous forests such as the birch-oak forest or the oak-hornbeam forest. Mixed deciduous-coniferous forests grow on a further 18%. They are clearly far removed from the natural forest vegetation. 10.1% of the stands are pure coniferous forests and thus very far removed from natural forest vegetation.

On the basis of the vegetation surveys, the mapped forest communities can be assigned to the ecosystem type classes according to Sect. 3.1.1 by comparing them with the reference values of the vegetation communities in the BERN database. This produces a map of the ecosystem type classes at the scale of the vegetation surveys.

Accordingly, 11 of the 78 ecosystem type classes are represented in the Kellerwald study area.

Using the indicator values of the mapped vegetation from the BERN database in conjunction with the reference profile data from the BÜK1000N soil overview map, the corresponding ecosystem service potentials could now be assigned to each class (Table 3.14).

Table 3.14 Ecosystem type classes in Kellerwald National Park and their ecosystem service potentials

Ecosystem type Class	Habitat services	Net primary biomass productivity	Carbon sequestration service	Area [km]²	Percentage share [%]
Suboceanic, moderately dry to fresh, poor larch forest	1	3	1	0.3	0.67
Suboceanic, moderately dry to fresh, moderately nutrient-rich beech forest/ wood	1	3.5	2	4.11	9.14
Suboceanic, moderately dry to fresh, moderately nutrient-rich spruce forest	2	4	2	7.45	16.6
Suboceanic, moderately dry to fresh, moderately nutrient-rich fir-beech forest	4	4	3	31.7	70.4
Suboceanic, moist, moderately nutrient-rich oak-hornbeam forest	5	4	3	0.22	0.49
Suboceanic, moderately dry to fresh, nutrient-rich beech forest	4	4	4	0.48	1.07
Suboceanic, moist, nutrient-rich oak-hornbeam-ash wood	5	4	3.5	0.11	0.24
Suboceanic, moist, nutrient-rich black alder floodplain forest	5	3	3	0.22	0.49
Suboceanic, moderately dry to fresh, nutrient-rich beech forest	4	5	5	0.15	0.33
Suboceanic, moderately dry to fresh, nutrient-rich mountain elm-summer linden-blockwood trees	5	4	4	0.17	0.38
Suboceanic, dry, nutrient-rich, carbonate-rich sessile oak dry forest	5	3	4	0.12	0.27

3.2.2   Result Maps of the Assessed Ecosystem Services of the Forest in the Kellerwald National Park

The allocation of assessment points for the ecosystem services of the 11 Kellerwald ecosystem type classes follows the methodological principle of transferring the integrated assessment points of the ecosystem services shown in Fig. 3.8.

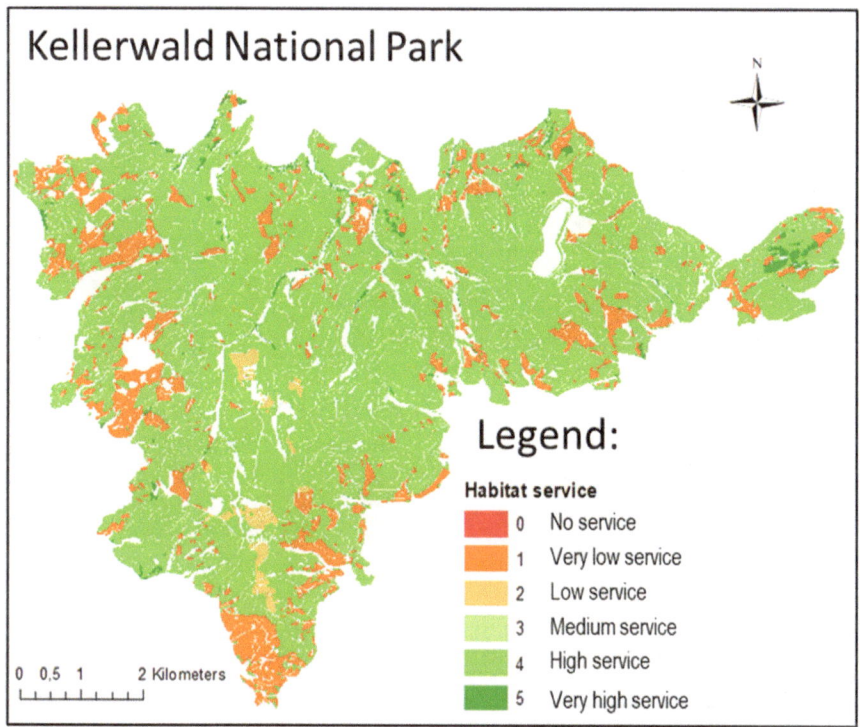

Fig. 3.15 Map of the assessment points for the habitat services of the 11 Kellerwald ecosystem type classes. (Source: Authors' own illustration)

The resulting maps are shown in the following Figs. 3.15, 3.16 and 3.17.

The distribution of the areas with the different assessment of their ecosystem services shows the high ecological importance of the national park (Table 3.15).

3.3 Regional Mapping

3.3.1 Modelling Approach and Data Basis for the Federal State of Saxony (Germany)

Schlutow and Gemballa (n.d.) carried out a project on the adaptation of the climate structure and the leading forest communities to climate change in the Free State of Saxony. This project resulted from the task of the "Forest Strategy 2050 for the Free State of Saxony", which sets out how forests and forestry in the Free State of Saxony must be organised in the middle of the 21st century in order to be able to meet the "current and future challenges with concrete proposals for action" (SMUL 2013).

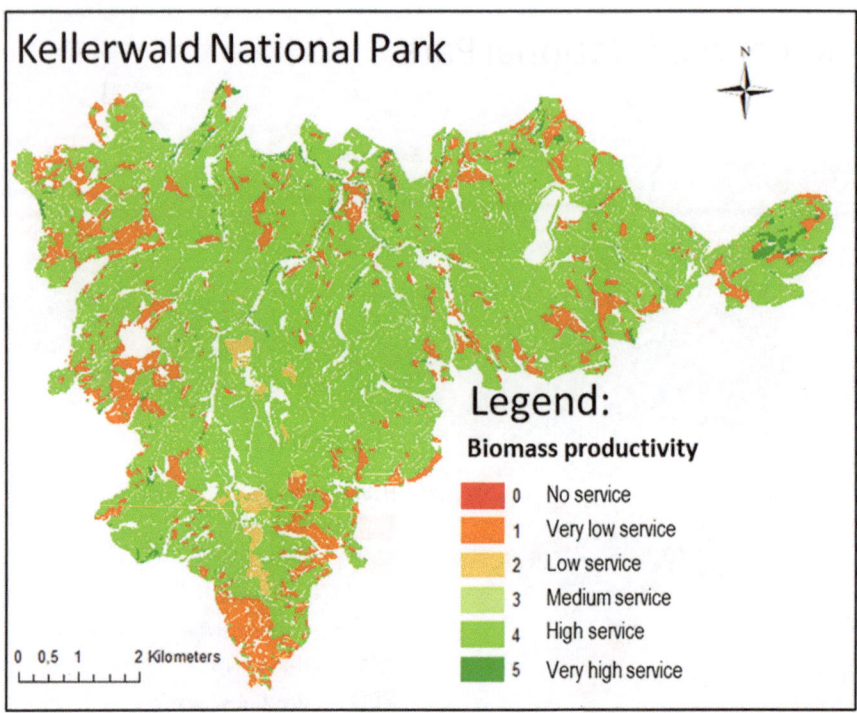

Fig. 3.16 Map of the evaluation points for the net primary biomass productivity of the 11 Kellerwald ecosystem type classes. (Source: Authors' own illustration)

The natural adaptation of existing tree species or the immigration of adapted, previously non-native tree species cannot keep up with the speed of climate change. In order to accelerate the adaptation process, the forest reorganisation already underway in the Free State of Saxony has been intensified to create site-appropriate, vital mixed forests. The identification of lead wood communities for all existing or expected climatic conditions in Saxony serves as a basis for planning.

By definition, the "lead forest community" is the plant community that is evolutionarily best adapted to the given site and climatic conditions, that has developed a dynamically stable competitive balance between the plant populations and their associated fauna and that can always regenerate itself in the event of disturbance. The indicator forest communities are therefore natural and semi-natural forest communities as documented in the BERN database and assigned to their preferred site and climate parameters (Schlutow et al. 2024).

For the identification of vital, self-regenerating and ecologically efficient indicator forest communities under the conditions of future climate change in Saxony, those near-natural forest communities are of particular interest that have developed and adapted, for example, in south-eastern Europe under the climatic conditions

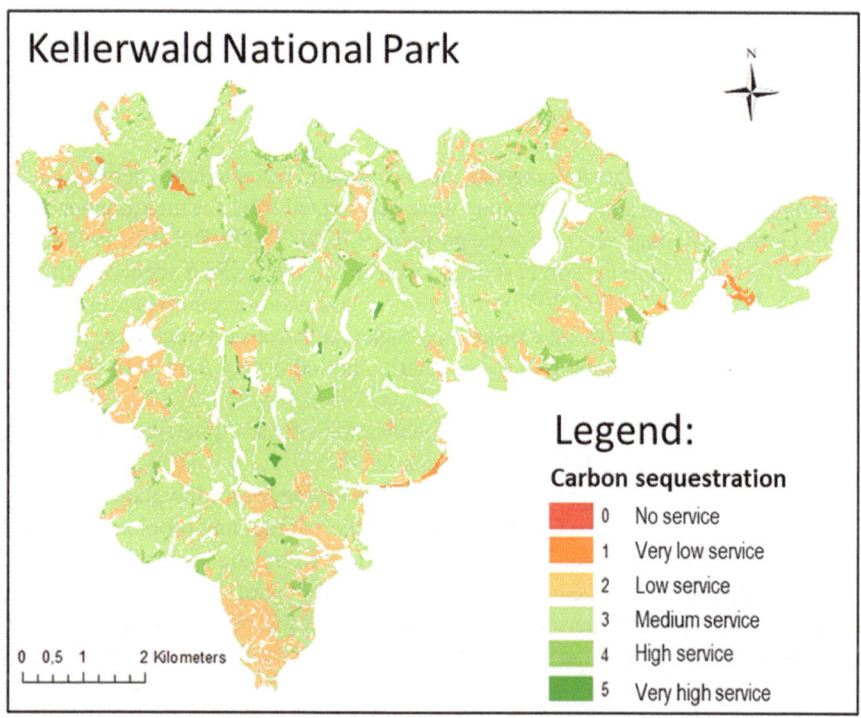

Fig. 3.17 Map of the assessment points for the carbon sequestration services of the 11 Kellerwald ecosystem type classes. (Source: Authors' own illustration)

Table 3.15 Distribution of the areas of the Kellerwald National Park with the different assessment of their ecosystem services

	Habitat services		Net biomass productivity		Carbon sequestration services	
	Rating grade	in [%]	Rating grade	in [%]	Rating grade	in [%]
No service	<0.5	–	< 0.5	0	0	0
Very little service	0.5–1.49	0.67	0.5–1.49	0	0.5–1.49	0.7
Low service	1.5–2.49	25.7	1.5–2.49	0	1.5–2.49	16.6
Intermediate service	2.5–3.49	0	2.5–3.49	1.4	2.5–3.49	80.5
High level of service	3.5–4.49	71.8	3.5–4.49	98.2	3.5–4.49	2
Very good service	4.5–5	1.9	4.5–5	0.3	4.5–5	0.3

that have prevailed there for centuries and are to be expected in Saxony in the future. The basis for this is the BERN database (Sect. 3.1.2).

The first step was to develop a climate classification for Saxony based on plant-physiologically relevant climate parameters. The demarcation of the climate classes should be able to move with the changing climate parameters in the future. The previous categorisation according to altitude levels therefore had to be abandoned. The two parameters - length of the vegetation period and climatic water balance in the vegetation period - are sufficient to establish a significant correlation with the occurrence of forest community groups (grouped according to main tree species) (Fig. 3.18).

The length of the vegetation period and the climatic water balance per vegetation month for the climate stages in Saxony were determined from the measurement data of the German Weather Service for the periods 1971–2000 (DWD 2012), 1991–2020 (DWD 2021) and from the forecast data according to the RCP8.5 scenario with the WEREX VI model (LfULG 2020) in a 1x1 km² grid according to two scenarios p1 (moderate case) and p2 (worst case). Daily mean temperature values were included in the calculation of the length of the vegetation period if they were > 10 °C on five consecutive days.

Figures 3.19 and 3.20 show the regionalisation of the climate classes for the periods 1991–2000 and 2041–2070 in scenario p1.

A direct link to the typical ecoclimatic parameter ranges for the plant communities in the BERN5 database was therefore possible without any problems. In order to allocate the indicator woodland communities to the climatic levels of Saxony, the possible ranges of existence of the communities with regard to the two ecoclimatic parameters of vegetation period length and climatic water balance were intersected with the corresponding value ranges of the Saxon climatic levels (Figs. 3.19, 3.20).

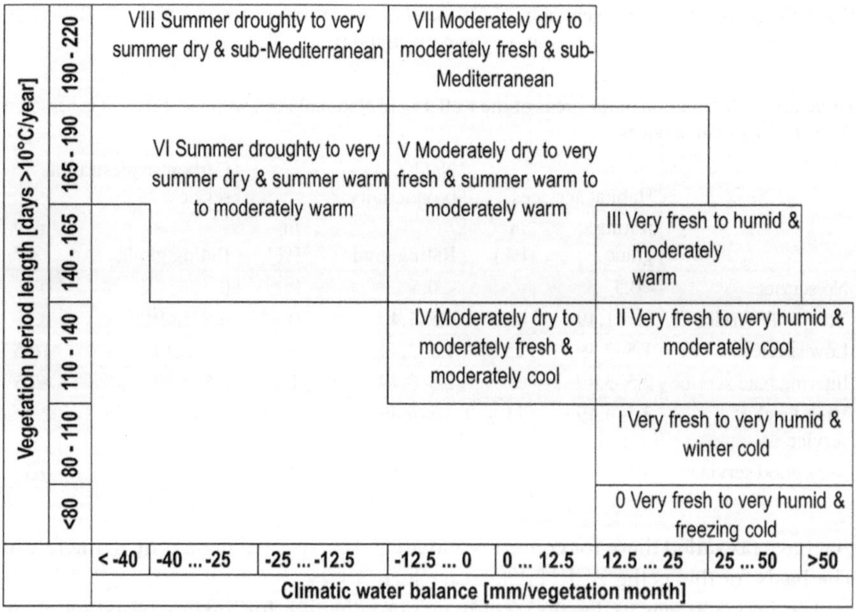

Fig. 3.18 Climate classes in Saxony, defined according to the length of the vegetation period and the climatic water balance (Source: Authors' own illustration)

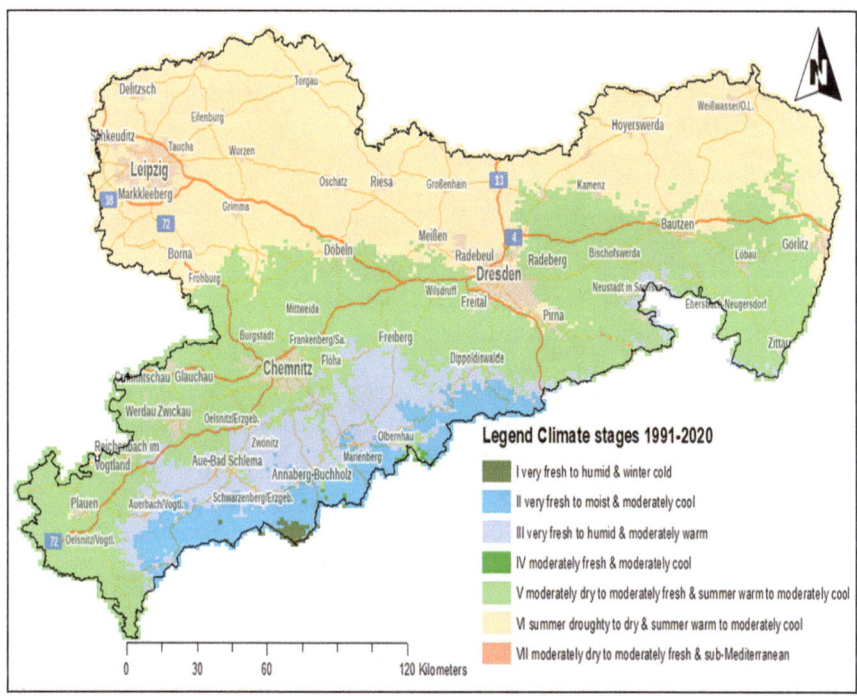

Fig. 3.19 Regionalisation of the climate stages for the periods 1991–2000. (Source: Authors' own illustration)

The regional distribution of soil types was based on the forest site map of Saxony at a scale of 1:10,000 (KWF 2022). The scheme from the forest site mapping according to SEA95 (Schulze 1998) was used to assign the C/N and base saturation value ranges for the plant communities of the BERN5 database to the site forms on the basis of the mapped nutrient thickness.

The 1555 different site types of the Saxon forest were combined with the respective relief and exposure variants (plain, shady and sunny slopes). The resulting 2724 combination types were then combined with the climate classes that currently prevail or are expected to prevail until 2070 according to the above-mentioned results of climate modelling. For these 22,000 current site combination types in Saxony, the lead forest communities were now assigned, whose community-typical value ranges for climate and site parameters largely correspond to the value ranges of the site and climate types of the Saxon forest site mapping. The 123 indicator forest communities selected for this purpose, each with the highest degree of viability under the current and future site conditions in Saxony, are to serve as target criteria for deriving the stocking target types for ecological forest conversion. These 123 indicator wood communities were summarised into 15 indicator wood groups (Table 3.16, Figs. 3.21, 3.22).

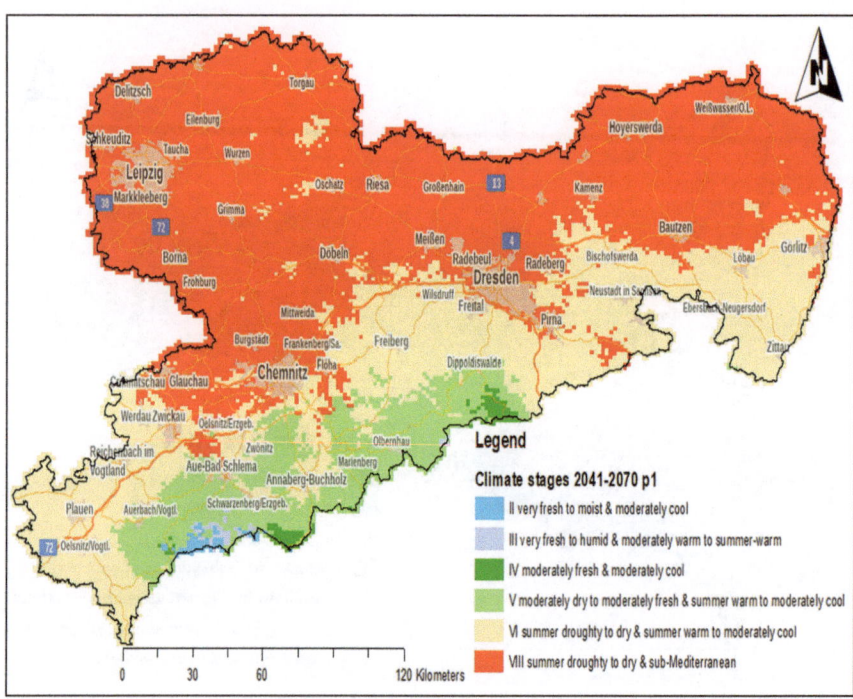

Fig. 3.20 Regionalisation of the climate stages for the period 2041–2070 according to scenario p1. (Source: Authors' own illustration)

In this way, the degree of correspondence between the ecological niche of a natural wood community and its structurally relevant tree species and the range of a climate/site type was determined. Since the ecological niche is defined as a fuzzy area (Roberts 1986), its existence outside the identified ecological niche is not completely excluded, even if the degree of possibility there is very low. It should also be noted that the ecological niche is usually narrower than the fundamental niche, as it is characterised by a natural competitive balance between the species populations at the site in addition to the fundamental niche. This means that individual tree species in pure stands created by forestry can also survive well outside their ecological niche, but their ecological functionality (e.g., habitat function) and resistance (e.g., to diseases and pests) is then significantly reduced.

Based on these findings by Schlutow and Gemballa (n.d.), assessment points could be awarded for ecosystem services. Since the semi-natural plant community serves as a summary indicator for the sites it colonises, particularly with regard to soil and climate parameters, an assessment of ecosystem services can be linked to the plant communities. The following assessment points are assigned to the possible forest communities in Saxony under current and future conditions for the 3 ecosystem services analysed in depth Table 3.16).

Table 3.16 Assessment points for the three in-depth ecosystem services of the possible forest communities in Saxony under current and future conditions [assessable points: 0…5]

Leitholz communities for Saxony	Leading wood group	Habitat service	Biomass productivity	Sequestration of carbon
Bromo-Carpinetum betuli HOFMANN 1968	(English oak) Hornbeam forest	4.6	3.1	2.2
Galio-Carpinetum betuli OBERDORFER 1957	(English oak) Hornbeam forest	4.6	3.4	3.5
Polytricho-Carpinetum betuli SCAMONI 1959	(English oak) Hornbeam forest	4.6	3.1	2.2
Primulo veris-Carpinetum betuli NEUHÄUSL et NEUHÄUSLOVA-NOVOTNA 1964	(English oak) Hornbeam forest	4.6	4.0	3.6
Querco roboris-Carpinetum betuli TX. 1937	(English oak) Hornbeam forest	4.6	4.0	3.6
Stachyo-Carpinetum betuli TÜXEN 1930	(English oak) Hornbeam forest	4.6	4.0	3.6
Stellario holosteae-Carpinetum betuli HARTMANN 1959	(English oak) Hornbeam forest	4.6	4.0	3.6
Tilio cordatae-Carpinetum betuli TRACZYK 1962	(English oak) Hornbeam forest	4.6	4.0	3.6
Aceri platanoides-Tilietum platyphylli WINTERHOFF 1962	Ash-Elm-Birchwood	4.6	4.3	3.9
Adoxo-Aceretum pseudoplatani PASSARGE 1959	Ash-Elm-Birchwood	4.6	4.0	3.5
Carici albae-Tilietum cordatae MÜLLER et GöRS 1958	Ash-Elm-Birchwood	4.6	3.6	3.8
Carici remotae-Fraxinetum excelsi W. KOCH 1926 ex FAB. 1936	Ash-Elm-Birchwood	4.6	4.2	4.7
Carpino betuli-Ulmetum carpinifoliae typicum PASS. 1953	Ash-Elm-Birchwood	4.6	3.6	3.8
Carpino betuli-Ulmetum scabrae HOFM. 1960	Ash-Elm-Birchwood	4.6	3.6	3.8
Fraxino excelsi-Aceretum pseudoplatani TÜXEN 1937	Ash-Elm-Birchwood	4.6	4.0	3.5
Luzulo luzuloides-Tilietum cordatae GRABHERR et MUCINA 1989	Ash-Elm-Birchwood	4.6	3.9	3.6
Pruno-Fraxinetum excelsi OBERDORFER 1953	Ash-Elm-Birchwood	4.6	3.7	4.7
Ulmo glabrae-Aceretum pseudoplatani TRAUTMANN 1952	Ash-Elm-Birchwood	4.6	3.6	3.8
Athyrio-Alnetum glutinosae TX. 1943	Alder wood	4.6	3.9	3.8

(continued)

Table 3.16 (continued)

Leitholz communities for Saxony	Leading wood group	Habitat service	Biomass productivity	Sequestration of carbon
Cardamino armarae-Alnetum glutinosae (MEIJER-DREES 1936) PASSARGE 1968	Alder wood	4.6	3.2	4.8
Carici elongatae-Alnetum glutinosae SCHWICKERATH 1933	Alder wood	4.6	3.2	4.8
Filipendulo-Alnetum LEMÉE 1937	Alder wood	4.6	3.2	4.8
Irido-Alnetum glutinosae DOING 1962	Alder wood	4.6	3.2	4.8
Stellario-Alnetum LOHMEYER 1957	Alder wood	4.6	3.2	4.8
Bazzanio-Piceetum (SCHMIDT et GAISBERG 1936) BR.-BL. et SISSINGH in BR.-BL. et al. 1939	Spruce wood	3.1	3.3	0.7
Calamagrostio variae-Piceetum SCHWEINGRUBER 1972	Spruce wood	1.3	3.8	2.9
Calamagrostio villosae-Piceetum VOLK 1939	Spruce wood	3.1	3.3	0.7
Carici albae-Piceetum MAYER et al. 1967	Spruce wood	1.3	3.8	2.9
Sphagno-Piceetum KUOCH 1954	Spruce wood	4.6	2.5	1.6
Vaccinio myrtilli-Piceetum TX. 1955	Spruce wood	1.3	3.8	2.9
Castaneo-Fagetum sylvatici MARINCEK 1980	Chestnut wood	4.0	3.8	2.7
Castaneo-Quercetum petraea HORVÁT 1963	Chestnut wood	4.0	2.6	2.5
Cladonio-Pinetum sylvestris PASS. 1956	Pine wood	4.6	1.5	0.5
Corynephoro-Pinetum sylvestris (JURASZEK 1928) HOFMANN 1964	Pine wood	4.6	1.5	0.5
Eriophoro-Pinetum sylvestris [HUECK 1925] HOFM. et PASS. 1968)	Pine wood	4.6	2.6	3.5
Festuco ovinae-Pinetum sylvestris JURASZEK 1928	Pine wood	4.6	1.5	0.5
Leucobryo-Pinetum sylvestris MATUSZ. 1962	Pine wood	4.6	1.5	0.5
Leucobryo-Pinetum variscum REINHOLD 1939	Pine wood	4.6	1.5	0.5
Peucedano-Pinetum sylvestris MATUCZ. 1962	Pine wood	4.6	1.5	0.5

(continued)

Table 3.16 (continued)

Leitholz communities for Saxony	Leading wood group	Habitat service	Biomass productivity	Sequestration of carbon
Pleurozio-Pinetum sylvestris KLEIST 1929	Pine wood	4.6	2.6	3.0
Pyrolo-Pinetum sylvestris (LIBBERT 1933) SCHMID 1936	Pine wood	3.5	2.6	2.3
Vaccinio myrtilli-Pinetum sylvaticae PASS. 1956	Pine wood	4.6	1.5	0.5
Vaccinio uliginosi-Pinetum sylvestris KLEIST 1929 em. MATUSZ. 1962	Pine wood	4.6	2.6	4.0
Corno-Quercetum (pubescentis-dalechampii) MATHÉ et KOVÁCS 1962	Mediterranean oak wood	3.0	2.6	2.5
Quercetum dalechampii-cerris SOÓ 1963	Mediterranean oak wood	3.0	2.6	2.5
Sorbo torminalis-Quercetum (dalechampii) SVOBODA ex BLAZKOVA 1962	Mediterranean oak wood	3.0	2.6	2.5
Eriophoro-Betuletum pubescentis HUECK 1925 em. PASSARGE 1968	Bog birch wood	4.6	2.6	3.5
Pleurozio-Betuletum pubescentis HUECK 1925 em. PASSARGE 1968	Bog birch wood	4.6	2.6	3.0
Sphagno-Betuletum pubescentis DOING 1962	Birch bog wood	4.6	2.6	4.0
Vaccinio uliginosi-Betuletum pubescentis LIBB. 1933	Bog birch wood	4.6	2.6	4.0
Carici piluliferae-Fagetum sylvatici PASS. 1956	Beech wood (oak)	4.4	3.8	2.7
Dicrano-Fagetum sylvatici PASS. et HOFM. 1965	Beech wood (oak)	4.6	2.8	1.0
Fraxino excelsi-Fagetum sylvatici SCAMONI 1956	Beech wood (oak)	5.0	4	3.5
Luzulo luzuloides-Fagetum sylvatici MEUSEL 1937	Beech wood (oak)	4.1	3.8	2.6
Maianthemo-Fagetum sylvatici PASS. 1959	Beech wood (oak)	4.4	3.3	2.6
Molinio-Fagetum sylvatici SCAM. 1959	Beech wood (oak)	5.0	2.5	3.5
Vaccinio myrtilli-Fagetum sylvatici PASS. 1965	Beech wood (oak)	4.6	2.9	1.2
Asperulo-Abieti-Fagetum sylvatici TH. MÜLLER 1964	Beech-fir wood	3.9	4.0	3.6

(continued)

Table 3.16   (continued)

Leitholz communities for Saxony	Leading wood group	Habitat service	Biomass productivity	Sequestration of carbon
Calamagrostio arundinaceae-Abieto-Fagetum sylvatici HARTM. et JAHN 1967	Beech-fir wood	4.4	3.4	1.5
Calamagrostio villosae-(Abieto-) Fagetum syvatici Mikuska 1972	Beech-fir wood	4.4	3.4	1.5
Luzulo-Abieto-Fagetum sylvatici HARTM. et JAHN 1967	Beech-fir wood	4.4	3.6	2.6
Asperulo-Fagetum sylvatici MAYER 1964	Beech wood	4.4	4.3	3.6
Impatiento-Fagetum sylvatici BARTSCH 1940	Beech wood	5.0	2.5	3.5
Melico-Fagetum sylvatici KNAPP em. 1942	Beech wood	3.9	3.8	3.5
Mercuriali-Fagetum sylvatici ((HARTMANN 1953) HOFMANN 1965	Beech wood	3.9	4.0	3.6
Vaccinio myrtilli-Fagetum sylvatici PASS. 1965	Beech wood	4.6	2.9	1.2
Aceri tartarici-Quercetum frainetto-pedunculiflorae STAJANOV 1955 em. ZÓLYOMI 1957	Oak trunk wood	3.0	2.6	2.5
Agrostio-Populetum tremulae PASS. and HOFMANN 1964	Oak trunk wood	4.0	2.6	2.5
Agrostio-Quercetum roboris deschampsietosum PASS. 1953 em. SCHUB. 1995	Oak trunk wood	4.5	2.6	2.5
Betulo-Quercetum roboris molinietosum (TX. 1937) SCAMONI et PASSARGE 1959	Oak trunk wood	4.5	3.4	1.8
Cytiso nigricantis-Quercetum roboris OBERD. 1957	Oak trunk wood	4.6	3.2	4.7
Dicrano-Quercetum (roboris) PASS. 1963	Oak trunk wood	4.5	3.4	1.8
Dictamno-Quercetum (roboris) FÖRSTER 1968	Oak trunk wood	4.6	3.2	4.7
Lysimachio-Quercetum roboris SCAM. et PASSARGE 1959	Oak trunk wood	4.6	3.2	4.7
Potentillo albae-Quercetum petraeae-roboris LIBBERT 1933	Oak trunk wood	4.5	2.6	2.5
Querco-Ulmetum ISSLER 1953	Oak trunk wood	4.7	4.1	3.6
Sambuco-Quercetum roboris HOFMANN 1965	Oak trunk wood	4.5	3.7	4.5
Vaccinio vitis-ideae-Quercetum (roboris) OBERD. (1957) 1992	Oak trunk wood	4.6	2.4	1.6

(continued)

Table 3.16 (continued)

Leitholz communities for Saxony	Leading wood group	Habitat service	Biomass productivity	Sequestration of carbon
Violo-Quercetum roboris TÜXEN et DIEMONT 1937 deschampsietosum PASS. 1953 em. SCHUB. 1995	Oak trunk wood	4.5	3.4	1.8
Adenostylo glabrae-Abietetum MAYER 1969	Fir (spruce) wood	4.4	3.3	2.0
Equiseto sylvatici-Abietetum albae MOOR 1952	Fir (spruce) wood	4.4	3.3	2.0
Vaccinio-Abietetum OBERDORFER 1957	Fir (spruce) wood	4.3	2.7	1.4
Betulo-Quercetum petraeae (GAUME 1924) TX. 1937	Sessile oak wood	4.5	3.4	1.8
Genisto tinctoriae-Quercetum (petraea) KLIKA 1932	Sessile oak wood	4.5	3.4	1.8
Holco mollis-Quercetum (robori-petraeae) LEMÉE 1937 corr. et em. OBERD. 1992	Sessile oak wood	4.6	3.2	2.6
Salicetum albae ISSLER 1926	Willow grove	4.6	3.7	3.9
Salicetum purpureae WENDELB.-ZELINKA 1952	Willow grove	4.6	3.7	3.9

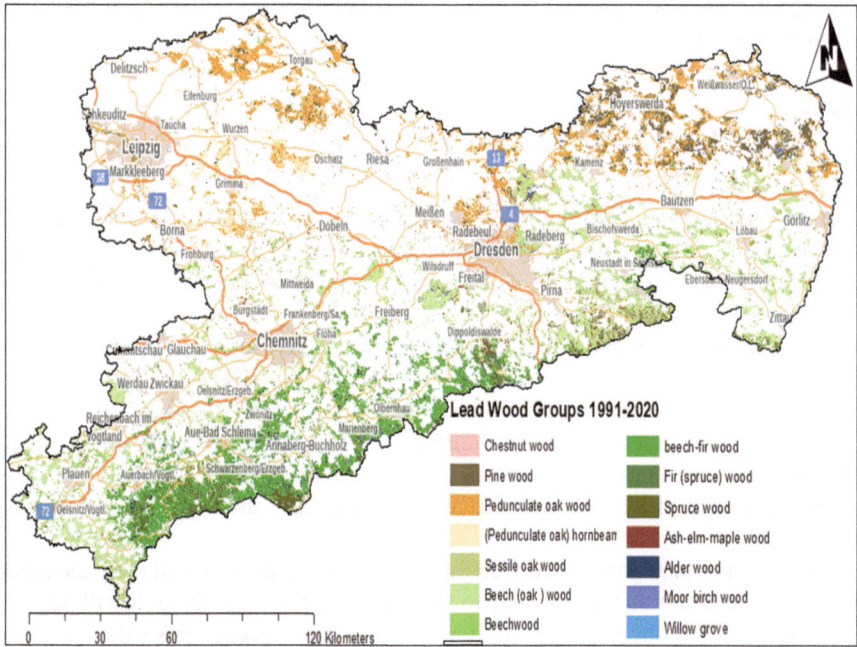

Fig. 3.21 Leading wood groups with the best vigour and resistance in the period 1991–2000. (Source: Authors' own illustration)

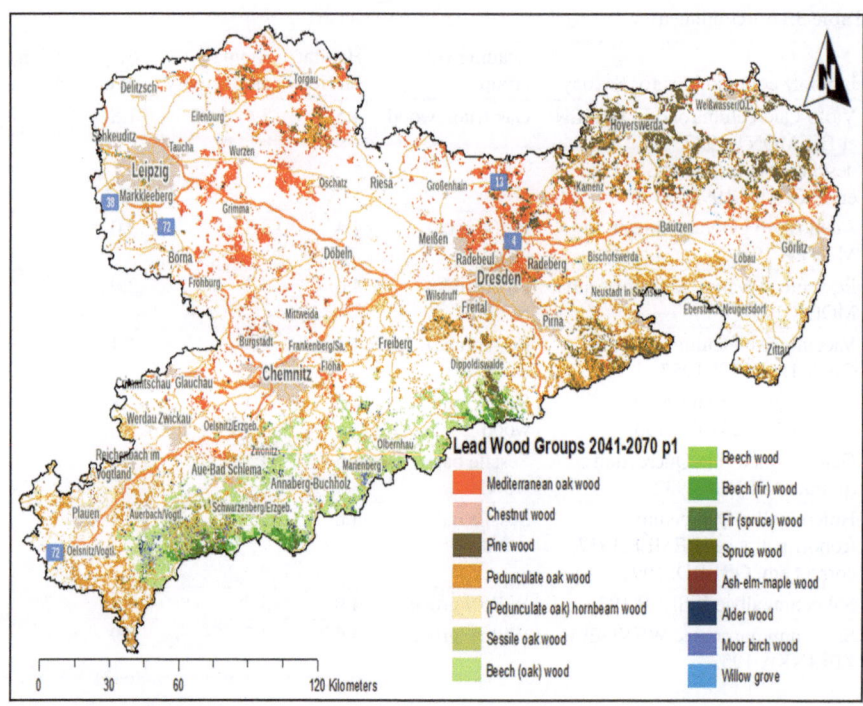

Fig. 3.22 Leading wood groups with the best vitality and resilience in the period 2041–2070 according to scenario p1. (Source: Authors' own illustration)

3.3.2   Result Maps of the Assessed Forest Ecosystem Services Under Climate Change in Saxony

The regionalised results of the rule-based assessment of the three in-depth ecosystem services for forests in Saxony in the periods 1991–2000 and 2041–2070 according to the climate projection RCP8.5 (p1) are shown in Figs. 3.23, 3.24, 3.25, 3.26, 3.27 and 3.28.

The comparison of the area shares [%] of the Saxon forests in the assessment results.

Table 3.17) for habitat services between the periods 1991–2020 and 2041–2070 according to the climate projection RCP8.5 (p1) shows that the values will decrease in more than half of Saxony's forests (North Saxon Lowlands). In future, the climatic conditions there will be extremely warmer and drier during the vegetation period.

On the other hand, habitat services are increasing in the hilly and montane areas in the southern part of Saxony. Due to the climatic conditions there, it will be possible in future to convert purely coniferous forests into mixed forests that can provide a better habitat function.

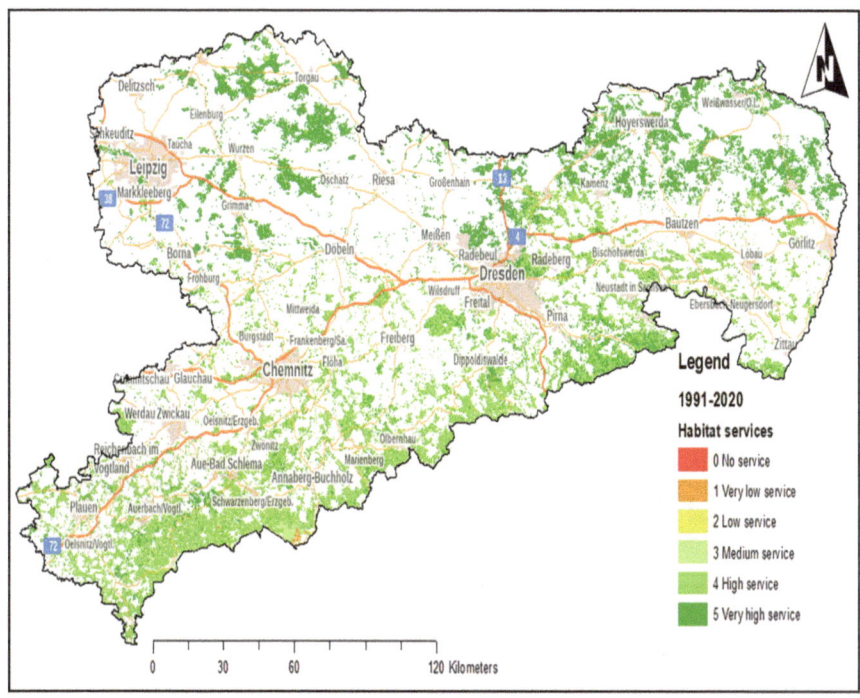

Fig. 3.23 Regionalised results of the rule-based assessment of habitat services in the period 1991–2000. (Source: Authors' own illustration)

Biomass production increases slightly in regions where temperature was a limiting factor in the growing season. In the subcontinental lowland regions, increasing drought will reduce biomass productivity.

The binding of carbon in the organic soil layer will decrease almost everywhere in Saxony in the future. The rising temperatures are causing a significantly faster mineralisation of the organic litter in the forest.

3.4 Location-Specific Rule-Based Classification of Ecosystem Services

3.4.1 Dynamic Modelling Approach and Data Basis for an LTER Area

The effects of acidifying and eutrophying air pollution and climate change on biodiversity have already been investigated using the example of a site in north-east Germany, which is integrated into the German part of the INTERNATIONAL LONG TERM ECOLOGICAL RESEARCH network (https://www.ilter.network/;

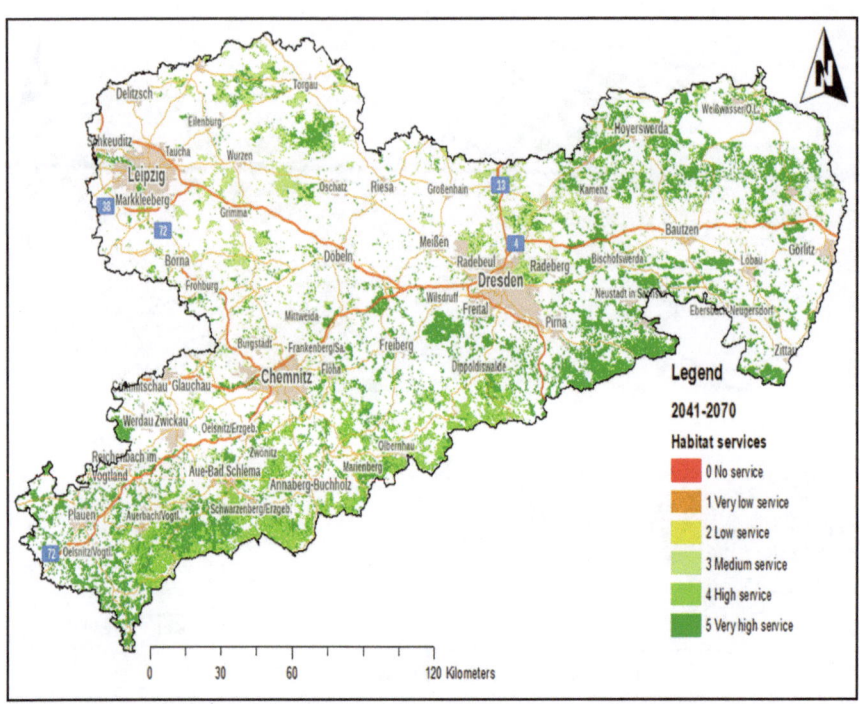

Fig. 3.24 Regionalised results of the rule-based assessment of habitat services in the period 2041–2070 according to the climate projection RCP8.5 (p1). (Source: Authors' own illustration)

Schlutow et al. 2015, Holmberg et al. 2018). For this purpose, data time series of atmospheric sulphur and nitrogen deposition from 1880 to 2000 were taken from EMEP (2018). The deposition data were scaled to the deposition data of the LTER site (Kranenburg n.d.; Figs. 3.29, 3.30). The deposition of nitrogen from 2017 onwards was set at a constant 15 kg N ha^{-1} a $^{-1}$.

The climate data for the 1950–2020 time series comes from the German Weather Service (DWD 2021). The time series from 2021–2070 were modelled by the authors according to RCP 8.5. The climate drivers were derived from the STAR II (Orlowsky et al. 2008) dataset (Figs. 3.31, 3.32).

The data on the effects of air pollution and climate change were fed into the geochemical model VSD+ (Bonten et al. 2016), which was parameterised for the local conditions of the LTER site based on the site description (Schulte-Bisping and Beese 2016). The results of VSD+ are time series of soil pH, base saturation and carbon-nitrogen ratio (C/N). The climatic conditions from the STAR II dataset range from 1900 to 2060 under different emission scenarios. Only the RCP 8.5 scenario was used as an application example. The resulting time axis from 1900 to 2060 for soil chemical factors is shown in Fig. 3.33.

Four selected time points from the time series of the driving parameters used in the BERN model (Schlutow et al. 2024) to calculate the development of the

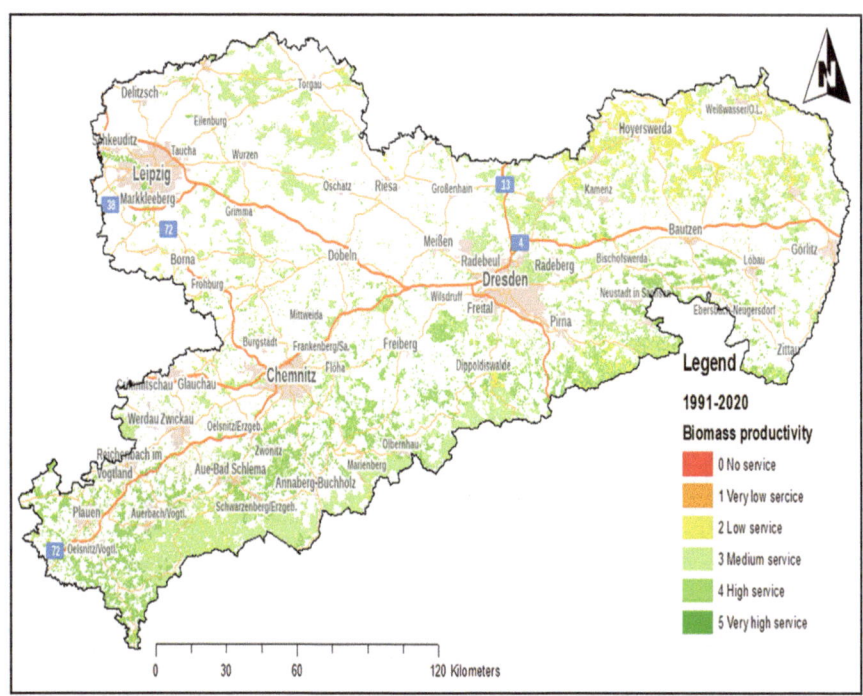

Fig. 3.25 Regionalised results of the rule-based assessment of biomass productivity in the period 1991–2000. (Source: Authors' own illustration)

viability of forest communities under changing climate and soil parameters are shown in Table 3.18.

Using the BERN model and the data set, the possibility of the existence of the plant community changes (Fig. 3.34).

Based on these results from Holmberg et al. (2018), the assessment of ecosystem services could be carried out (Table 3.19). The site characteristics from the combination of soil and climate parameters could be classified accordingly (Sect. 3.1.1). The former, current and future potential forest community is an indicator for this ecosystem type class.

However, the results should be evaluated with caution. The modelling of possible future communities in response to climate change and/or the effects of pollution can only be as realistic as the climate and deposition projections are realistic. As soon as new findings lead to refined climate and deposition models, the modelling with the BERN model must also be updated.

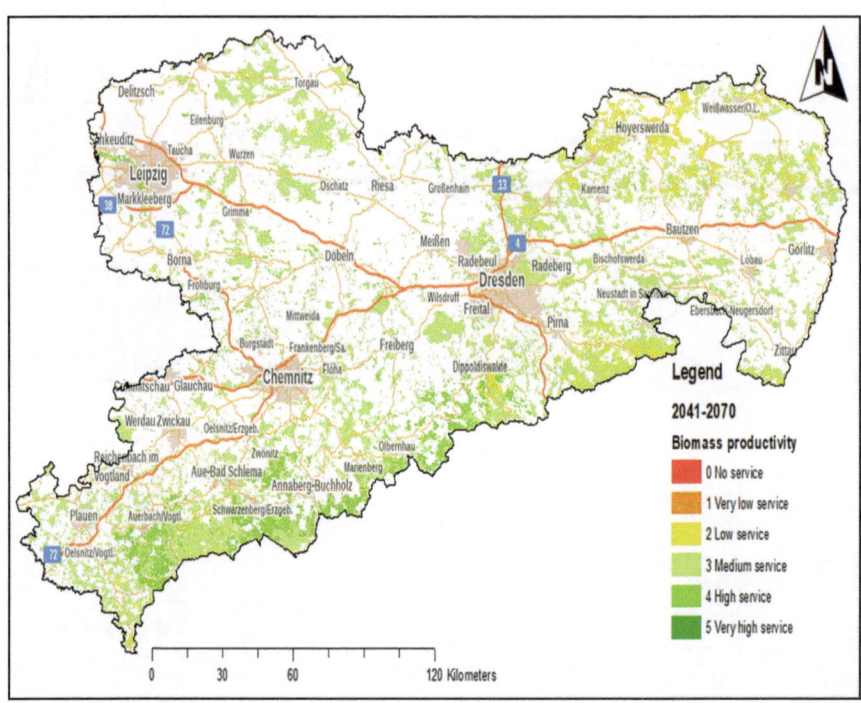

Fig. 3.26 Regionalised results of the rule-based assessment of biomass productivity in the period 2041–2070 according to the climate projection RCP8.5 (p1). (Source: Authors' own illustration)

3.4.2   Results of the Site-Specific, Rule-Based Classification of Ecosystem Services Under the Influence of Climate Change and Air Pollution

The development of ecosystem services at a site from the former natural deciduous forest vegetation through land-use change to coniferous forest to natural forest conversion back to deciduous forest is shown in the Figs. 3.35, 3.36 and 3.37.

The habitat supply decreased from 1900 to 2040 due to the conversion to an unstructured, site-untypical coniferous forest. The effect of the increasing biomass production achieved by this conversion has materialised. Carbon sequestration has decreased in recent decades due to rising autumn temperatures and will continue to decrease in the future.

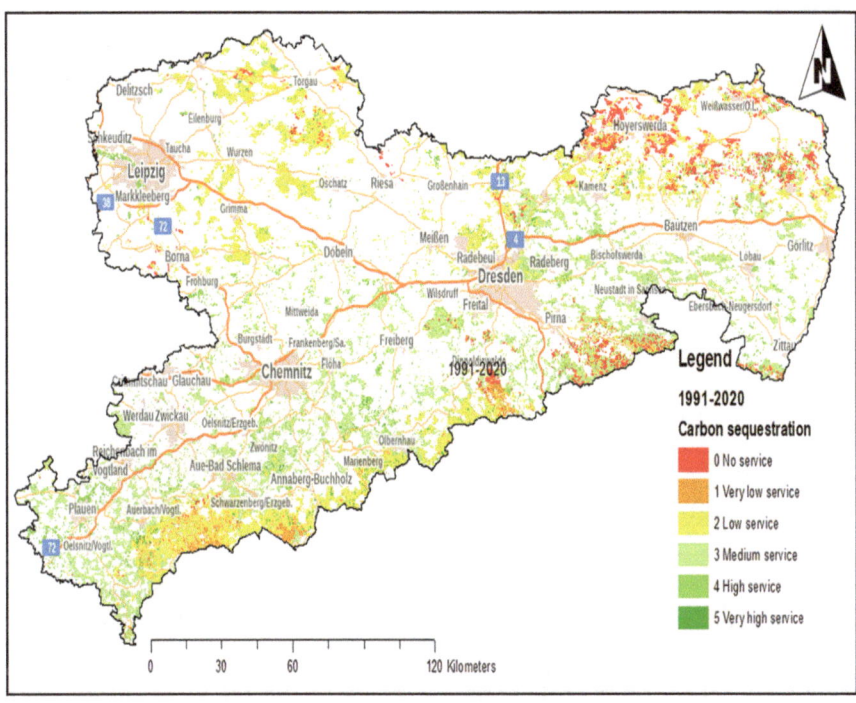

Fig. 3.27 Regionalised results of the rule-based assessment of carbon sequestration in the period 1991–2000. (Source: Authors' own depiction)

3.5 Rule-Based Classification of Open Land Habitats Under Different Land Uses

The rules for evaluating ecological services described in Sect. 2 can also be applied to open land biotopes without further ado.

However, nationwide mapping is difficult, as the mostly very small-scale biotopes cannot be represented in this study.

For this reason, an evaluation in the form of tables is provided below.

(a) Habitat service

The results after applying the rules for assessing the habitat service of some biotope types are shown in Table 3.20.

The data for the assessment comes from the BERN database. In addition to near-natural, more or less undisturbed plant communities, you will also find plant communities of open land, in particular meadow, pasture, moorland and heathland communities as well as arable weed communities with the corresponding information and for forest communities.

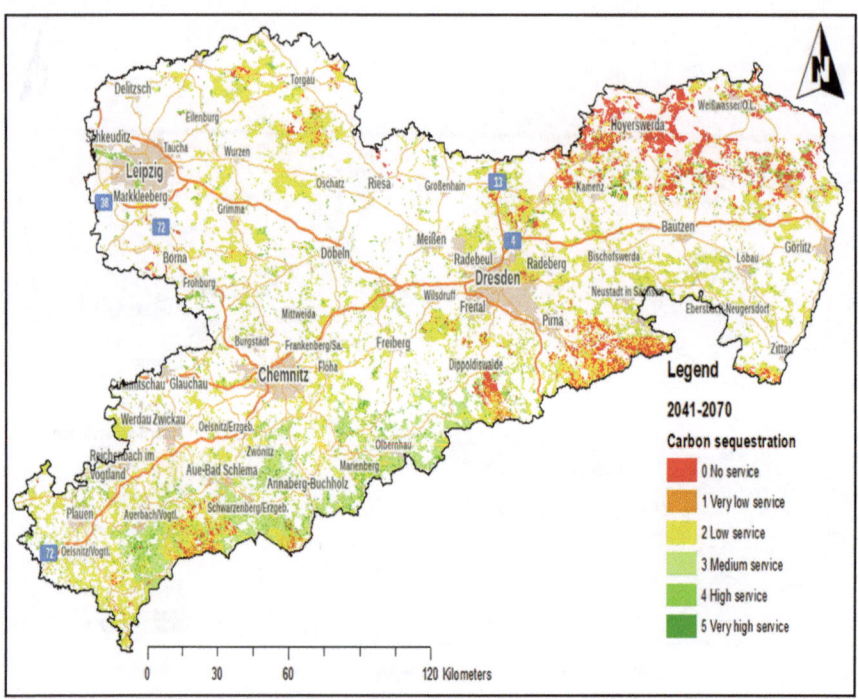

Fig. 3.28 Regionalised results of the rule-based assessment of carbon sequestration in the period 2041–2070 according to the climate projection RCP8.5 (p1). (Source: Authors' own illustration)

(b) Biomass primary production

A distinction is made between open land areas that are not utilised and those that are regularly used (mown meadows and pastures) or on which maintenance measures are carried out (weeding, removal of unwanted woody growth, mowing of reeds and reed beds, etc.).

The productivity of regularly used grassland depends mainly on the amount of fertiliser applied, so that a generalisation is only possible for unfertilised, extensively used grassland biotopes.

The net biomass productivity depends on the biomass production potential of the respective site. The more fertile the site, the higher the utilisation, which means that a higher extraction rate must be assumed (Table 3.21). However, the upper range limit ($E_{max(phyto)}$) does not indicate the physiologically maximum possible dry matter yield, but rather the minimum biomass yield compatible with the site on the most fertile typical soils of the respective vegetation type in a favourable climate. A minimum yield that can also be achieved under unfavourable conditions is also calculated theoretically ($E_{min(phyto)}$).

Table 3.17 Area shares [%] of Saxon forests in the assessment results for the three in-depth ecosystem services

Result	Habitat			Biomass			Carbon		
	1991–2020	2041–2070	Trend	1991–2020	2041–2070	Trend	1991–2020	2041–2070	Trend
0	0	0	↑	0	0	→	6	10	←
1	0	0		10	11		3	1	
2	2	20		60	75		38	64	
3	55	19	→	29	15	←	42	17	
4	43	61		0	0		9	7	
5	0	0		0	0		2	1	

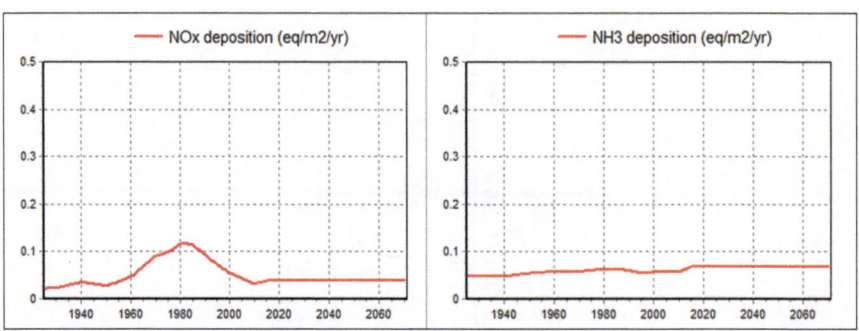

Fig. 3.29 Time series of deposition of NO$_x$ (left) and NH$_3$ (right). (Sources: Authors' own representation based on (**a**) 1920–2000: EMEP 2018; (**b**) Kranenburg n.d.; (**c**) 2018–2070: set to constant)

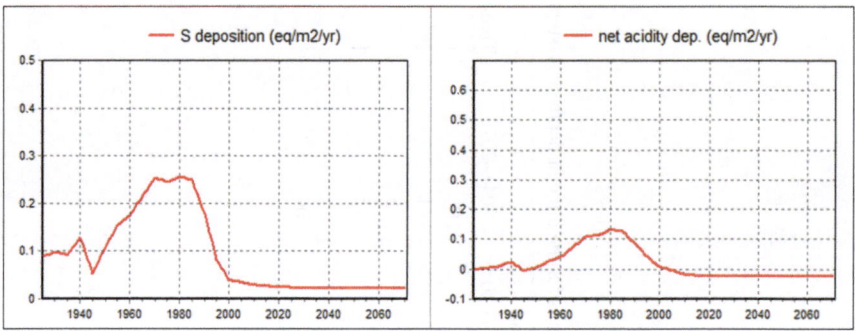

Fig. 3.30 Time series of the deposition of sulphur (left) and sulphur + nitrogen (right). (Sources: Own illustration based on (**a**) 1920–2000: EMEP 2018; (**b**) Kranenburg n.d.; (**c**) 2018–2070: set to constant)

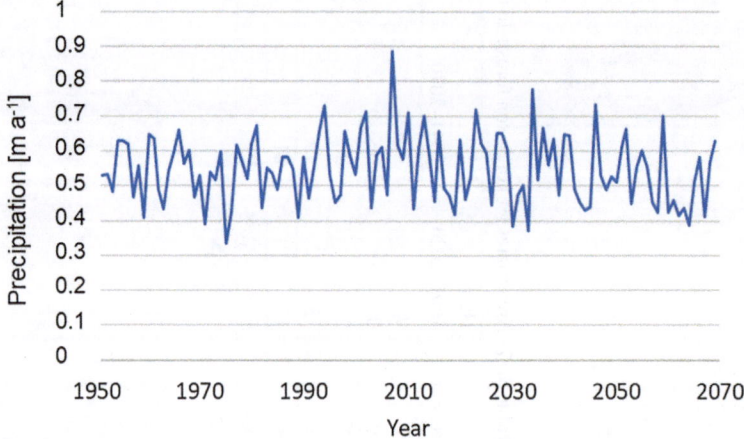

Fig. 3.31 Time series of precipitation for site-specific modelling. (Sources: Own representation based on (**a**) 1950–2020: DWD 2021; 2021–2070: Modelled by the authors according to RCP 8.5

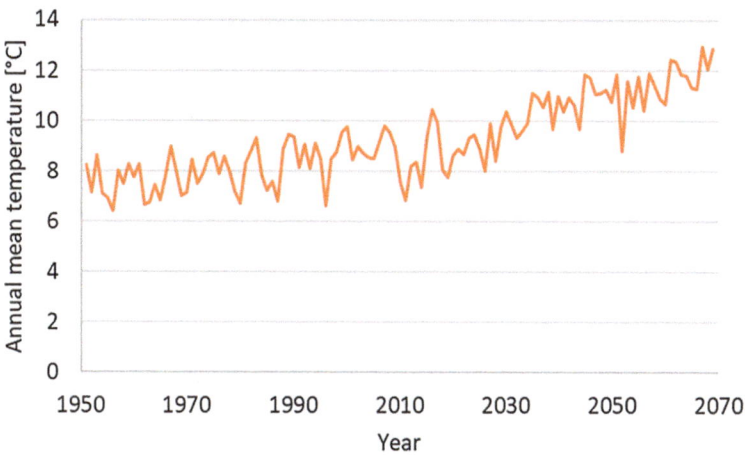

Fig. 3.32 Time series of annual mean temperature for site-specific modelling (sources: own representation based on (**a**) 1950–2020: DWD 2021; 2021–2070: modelled by the authors according to RCP 8.5)

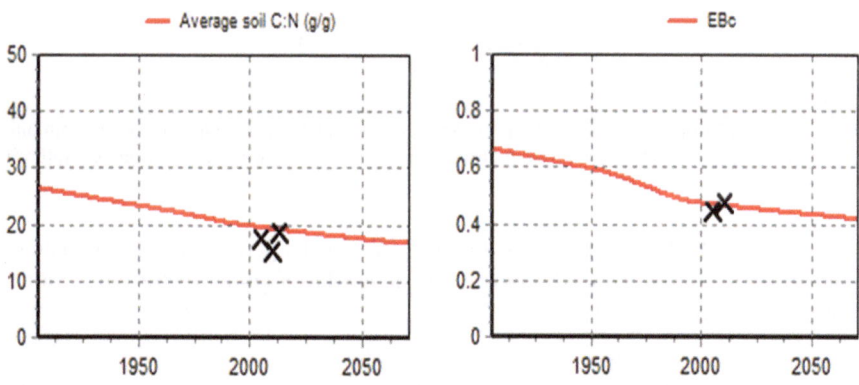

Fig. 3.33 Soil chemical driver variables for the LTER site (E_{Bc} = base saturation [mol_C /mol_C], C:N = ratio of carbon to nitrogen in g/g, black crosses indicate measured data). (Sources: Authors' own presentation)

Table 3.18 Time series of parameters as drivers for the BERN model

Parameters of the LTER location	1920	1985	2009	2040 (RCP8.5)	2070 (RCP8.5)
T (°C)	7.1	7.6	8.9	9.6	12.9
Precipitation (mm a)-[1]	530	582	618	473	441
Water content in rooted soil layers [m^3 m]-[3]	0,11	0,18	0.26	0.13	0.09
Climatic water balance (mm/veg. month)	−11.7	−9.5	−12.5	−30.6	−41.8
Growing season (d a)-[1]	165	165	176	185	190
Sdep (keq ha^{-1} a)-[1]	0.9	2.5	0.31	0.31	0.31
Ndep (keq ha^{-1} a)-[1]	0.8	1.8	0.9	1.07	1.07
BS (%)	63	44	40	35	31
C/N	25	21	19	18	17

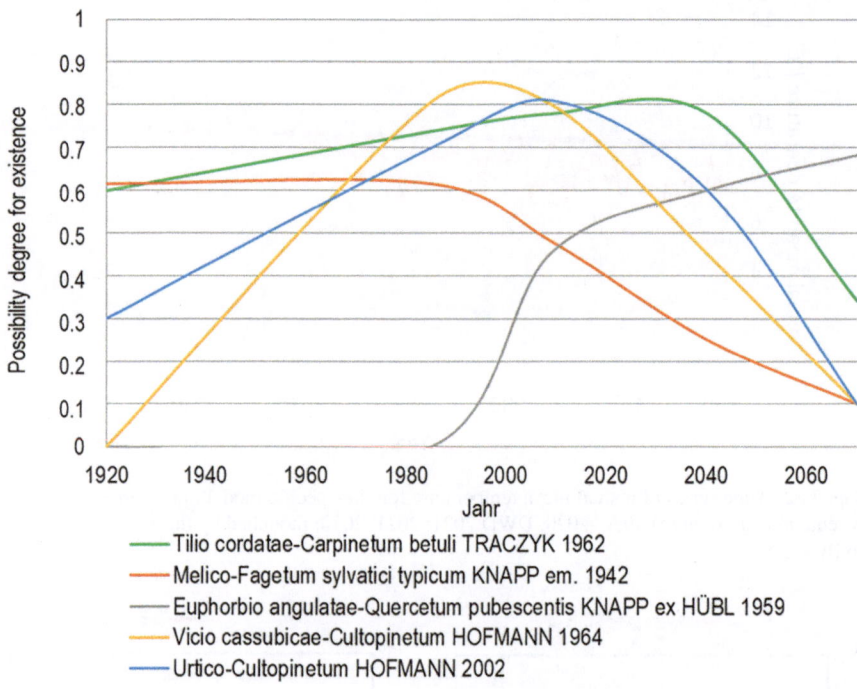

Tilio cordatae-Carpinetum betuli TRACZYK 1962
Melico-Fagetum sylvatici typicum KNAPP em. 1942
Euphorbio angulatae-Quercetum pubescentis KNAPP ex HÜBL 1959
Vicio cassubicae-Cultopinetum HOFMANN 1964
Urtico-Cultopinetum HOFMANN 2002

Fig. 3.34 Results of the BERN model for calculating the potential existence of forest communities at the LTER site with the current anthropogenic mixed beech-pine stand. (Source: Author's own illustration)

In general, the following ratings (Table 3.22) for the periodically maintained or extensively utilised open land biotopes can be used as a guideline for typical locations of the biotope types:

(c) Carbon sequestration service

Application of the rules in Sect. 2 leads to a result that was to be expected for the relevant open land biotopes (Table 3.23). Wetlands and peatlands have particularly high carbon storage capacities. However, heathlands also accumulate organically bound carbon due to the strong and thus slow decomposition of the woody heathland biomass and its high proportion of residues after grazing.

Table 3.19 Time table of the former, current and expected future forest community and their assessment of ecosystem services

Year	Class ecosystem	Wood/forest community	Habitat service	Biomass productivity	Sequestration of carbon
<1920	Temperate suboceanic to temperate subcontinental climate, dry, vigorous, nutritious, beech wood/forest	Melico-Fagetum sylvatici typicum KNAPP em. 1942	3.9	3.8	3.5
1985	Temperate suboceanic to temperate subcontinental climate, moderately dry to fresh, vigorous, nutrient-rich pine forest	Vicio cassubicae-Cultopinetum HOFMANN 1964	1.5	1.5	3
2009	Temperate suboceanic to temperate subcontinental climate, moderately dry to fresh, vigorous, nutrient-rich pine forest	Urtico-Cultopinetum HOFMANN 2002	1	1.5	3
2040	Temperate suboceanic to temperate subcontinental climate, dry, vigorous, nutritious lime-hornbeam wood	Tilio cordatae-Carpinetum betuli TRACZYK 1962	4.6	4	3.6
2070	Temperate Central European to subcontinental climate, dry, moderately nutritious oak wood	Euphorbio angulatae-Quercetum pubescentis KNAPP ex HÜBL 1959	3	2.6	2.5

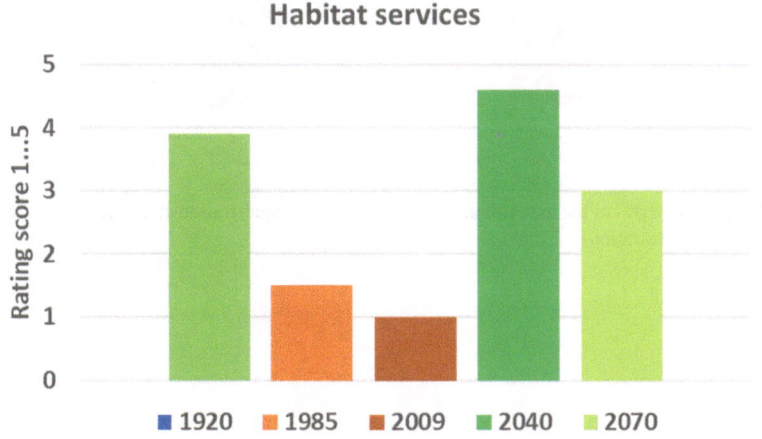

Fig. 3.35 Time series of assessment results for habitat services at the LTER site. (Source: Author's own illustration)

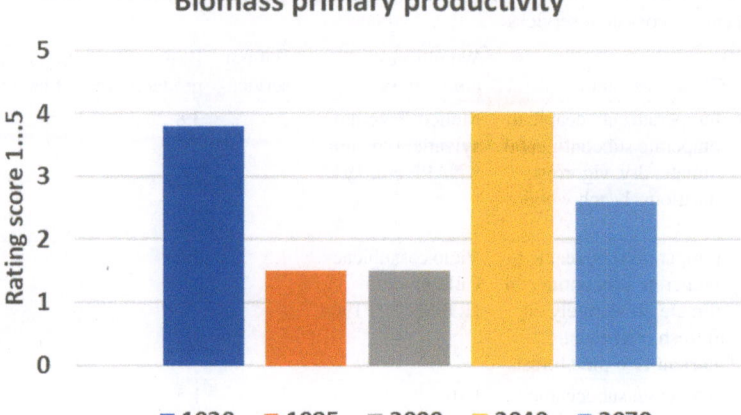

Fig. 3.36 Time series of the evaluation results for biomass productivity at the LTER site. (Source: Author's own illustration)

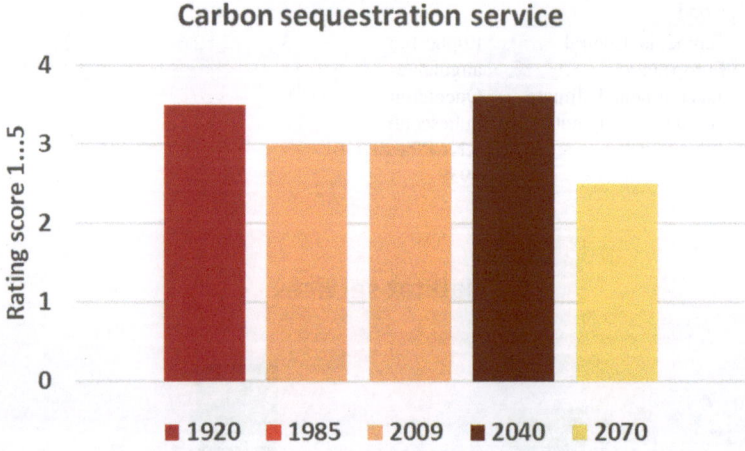

Fig. 3.37 Time series of the assessment results for carbon sequestration at the LTER site. (Source: Author's own illustration)

Table 3.20 Rule-based assessment points for the habitat service of open land biotopes

Biotope type	Hemeroby	Recoverability	Need for protection	Maturity level	Botope network	Habitat value for fauna			Completeness of the composition of the characteristic plant species	Integrative assessment score
						Mammels	Birds	Herpetofauna		
Near-natural stream, ditch	4.5	4.0	5.0	4.0	5.0	4.0	5.0	5.0	5.0	4.8
Sedge and reed swamps, peat bogs (undisturbed)	5.0	4.0	5.0	5.0	2.5	5.0	5.0	5.0	5.0	4.9
Willow scrub in moist locations	4.5	4.0	5.0	4.0	2.5	4.0	5.0	5.0	5.0	4.8
Reed communities in standing and flowing waters (naturally eutrophic)	4.5	4.0	5.0	4.0	2.5	4.0	5.0	5.0	5.0	4.8
Small water bodies	4.5	4.0	5.0	4.0	2.5	4.0	5.0	5.0	5.0	4.8
Peat cutting	3.5	4.0	5.0	4.0	2.5	4.0	5.0	4.0	4.0	4.2
Deciduous shrubs for dry and warm locations	5.0	4.0	5.0	3.0	2.5	5.0	5.0	1.0	4.0	4.0
Great sedge meadows	3.5	2.5	5.0	3.0	2.5	5.0	5.0	5.0	5.0	4.7
Wet meadow	3.5	2.5	5.0	3.0	2.5	5.0	5.0	5.0	5.0	4.7
Moist tall herbaceous meadow	3.5	2.5	5.0	3.0	2.5	5.0	5.0	5.0	5.0	4.7
Backwaters with forested fringe	3.5	4.0	3.5	4.0	2.5	4.0	5.0	4.0	4.0	4.0
Ponds	3.5	4.0	3.5	4.0	2.5	4.0	5.0	4.0	4.0	4.0
Cemetery	4.5	4.0	3.5	4.0	2.5	2.0	5.0	2.0	4.0	3.7
Park	4.5	4.0	3.5	4.0	2.5	2.0	5.0	2.0	4.0	3.7

(continued)

Table 3.20 (continued)

Biotope type	Hemeroby	Recoverability	Need for protection	Maturity level	Botope network	Habitat value for fauna				Completeness of the composition of the characteristic plant species	Integrative assessment score
						Mammels	Birds	Herpetofauna			
Cairns with shrubs	2.5	2.5	5.0	3.0	2.5	4.0	4.0	3.0		4.0	3.9
Deciduous shrubs in fresh locations	4.5	2.5	3.5	4.0	2.5	4.0	4.5	2.0		3.0	3.2
Old solitary trees	4.5	2.5	3.5	4.0	2.5	4.0	4.5	2.0		3.0	3.2
Groups of trees, rows of trees	4.5	2.5	3.5	4.0	2.5	4.0	4.5	2.0		3.0	3.2
Woods, hedges	4.5	2.5	3.5	4.0	2.5	4.0	4.5	2.0		3.0	3.2
Orchard meadows, old fruit trees	1.5	2.5	5.0	3.0	2.5	4.0	4.5	2.0		3.0	3.3
Sandy dry grassland on acidic soils	3.5	0.0	5.0	3.0	2.5	1.0	2.0	3.0		5.0	3.8
Dry grassland on alk	3.5	0.0	5.0	3.0	2.5	1.0	2.0	3.0		5.0	3.8
Dry stone wall	3.0	2.5	3.5	2.0	2.5	1.0	2.0	3.0		5.0	3.7
Stream, ditch, semi-natural, temporarily dry	1.0	2.5	2.0	2.0	2.5	2.5	3.0	3.5		3.5	3.0
Dummy trees	1.5	4.0	5.0	3.0	2.5	1.0	2.0	1.0		2.5	2.6
Lakes (> 5 m water depth)	4.5	4.0	5.0	4.0	2.5	4.0	5.0	5.0		5.0	4.8
Coniferous forests, mixed forests with non-native tree species	1.0	4.0	3.5	3.0	2.5	5.0	5.0	4.0		2.5	3.2
Forests in dry locations	5.0	4.0	3.5	5.0	2.5	4.0	3.0	3.0		5.0	4.3

Clearing and reforestation, initial afforestation	1.5	4.0	2.0	3.0	0.5	2.0	2.0	1.0	1.5	1.8
Fresh meadow, fresh pasture	2.5	0.0	2.0	1.0	2.5	4.0	4.0	4.0	4.5	3.6
Herbaceous meadows in fresh locations	2.5	0.0	2.0	1.0	2.5	4.0	4.0	4.0	4.5	3.6
Ruderal meadows	2.0	0.0	2.0	3.0	0.5	4.0	4.0	4.0	3.0	2.9
Gardens and garden wastelands	1.5	2.5	2.0	1.0	0.5	1.0	4.0	3.0	2.5	2.4
Frequently used fields	1.5	2.5	2.0	1.0	0.5	1.0	4.0	3.0	2.5	2.4
Small housing estates with gardens	1.5	2.5	2.0	1.0	0.5	1.0	4.0	3.0	2.5	2.4
Development of weekend and holiday homes	1.0	0.0	0.0	1.0	0.5	0.5	0.5	0.5	2.0	1.2
Open, vegetation-free areas without visible, current anthropogenic use	0.0	0.0	0.5	0.0	2.0	2.0	2.0	2.0	0.0	0.6
Intensive fruit growing	1.0	0.0	2.0	1.0	2.5	1.0	2.0	1.0	1.5	1.4
Golf course	2.5	0.0	2.0	1.0	2.5	4.0	4.0	4.0	4.5	3.6
Village green with groves	3.5	4.0	5.0	4.0	2.5	4.0	5.0	4.0	4.0	4.2
Campsite	1.0	0.0	0.0	1.0	0.5	0.5	0.5	0.5	2.0	1.2
Intensive arable land, intensive grassland	1.0	0.0	2.0	1.0	0.5	1.0	1.0	0.5	1.5	1.3
Bathing spots	2.0	0.0	2.0	1.0	0.5	0.5	0.5	0.5	1.5	1.2
Sports field, outdoor pool	1.0	0.0	0.0	1.0	0.5	0.5	0.5	0.5	2.0	1.2

(continued)

Table 3.20 (continued)

Biotope type	Hemeroby	Recoverability	Need for protection	Maturity level	Botope network	Habitat value for fauna				Completeness of the composition of the characteristic plant species	Integrative assessment score
						Mammels	Birds	Herpetofauna			
Settlement centre, industrial and commercial area	0.0	0.0	0.0	0.0	0.0	0.5	1.0	0.5		0.5	0.4
Technical infrastructure, supply and disposal facilities	0.0	0.0	0.0	0.0	0.0	0.5	0.5	0.5		0.5	0.4
Waste, rubble and other debris	0.0	0.0	0.0	0.0	0.0	0.5	0.5	0.0		1.0	0.6
Traffic facilities	0.0	0.0	0.0	0.0	0.0	0.0	0.0	0.0		0.5	0.3
Piped ditch	0.0	0.0	0.0	0.0	0.0	0.0	0.0	0.0		0.5	0.3
Nurseries, commercial horticulture	1.0	0.0	2.0	1.0	2.5	1.0	2.0	1.0		1.5	1.4
Anthropogenically utilised special areas	1.0	0.0	0.0	1.0	0.5	0.5	0.5	0.5		2.0	1.2
Ruins (outside settlements)	1.0	0.0	0.0	1.0	0.5	0.5	0.5	0.5		2.0	1.2

Table 3.21 Range of yield potentials (dry matter DM) of the different vegetation types of the forest-free semi-natural ecosystems (Schlutow et al. 2021)

Type of vegetation	Average annual growth rates	
	[t dry matter ha^{-1} a^{-1}]	
	$E_{min\ (Phyto)}$	$E_{max(Phyto)}$
Grassland	0.65	1.5
Heath landscape	0.7	1.5
Dry calcareous grassland	0.8	1.4
Moist and marshy meadows	0.11	1.7
Floodplain meadows and alluvial forests	0.1	2.5
Fresh meadows/fresh pastures	0.8	1.6

Table 3.22 Rule-based assessment results for the net biomass productivity of open land biotopes

	Plant physiological production	Water balance of the soil	Nutrient balance	Structure of the floor	Climate-specific productivity	Integrative evaluation
Acidic dry grassland	1.4	1	1	3	4	1.5
Heath landscape	1.5	1	1	3	4	1.5
Dry calcareous grassland	1.5	1	3	4	4	2.1
Moist and marshy meadows	1.2	5	3	2.5	2.5	3.8
Floodplain meadows and alluvial forests	1.7	5	5	2.5	2.5	4.3
Fresh meadows/ fresh pastures	1.6	2.5	3	3	2.5	2.6

Table 3.23 Rule-based assessment points for the carbon sequestration service of open land biotopes

	Clay content	Mass of litter and crop residues	Yield potential of the soil	pH value in humus	Climate influence	Soil water contamination	Decomposition time	Integrative evaluation
Acidic dry grassland	0.1	1	1	1	4	1	1	1.4
Heath landscape	0.1	5	1	1	4	1	5	2.0
Dry calcareous grassland	5	1	3	5	4	1	1	3.2
Moist and marshy meadows	4	5	3	4	2.5	5	4	3.6
Floodplain meadows and alluvial forests	3	5	5	2.5	2.5	5	3	4.0
Fresh meadows/fresh pastures	2	3	3	2.5	2.5	2.5	2	2.8

References

Achermann B, Bobbink R (eds) (2003) Empirical critical loads for nitrogen. Proceedings of the Expert workshop in Berne 11–13 November 2002. Environmental Documentation No 164 - Air (Swiss Agency for the Environment, Forests and Landscape SAEFL) pp 305–311

AG Boden - Arbeitsgruppe Boden (2005) Bodenkundliche Kartieranleitung [Soil mapping guide]. 5. Aufl., Bundesanstalt für Geowissenschaften und Rohstoffe und den Geologischen Landesämtern der Bundesrepublik Deutschland (Ed.), Hannover

Balla S, Uhl R, Schlutow A, Lorentz H, Förster M, Becker C, Scheuschner Th, Kiebel A, Herzog W, Düring I, Lüttmann J, Müller-Pfannenstiel K (2013) Untersuchung und Bewertung von straßenverkehrsbedingten Nährstoffeinträgen in empfindliche Biotope. Hrsg.: BMVBS—Bundesministerium für Verkehr, Bauwesen und Städtebau. Endbericht zum FE-Vorhaben 84.0102/2009 im Auftrag der Bundesanstalt für Straßenwesen [Investigation and assessment of road traffic-induced nutrient inputs into sensitive biotopes. Final report on FE project 84.0102/2009 on behalf of the Federal Highway Research Institute], written by = Forschung Straßenbau und Straßenverkehrstechnik, Heft 1099, BMVBS Abteilung Straßenbau, Bonn. 362 pp

BGR (Bundesanstalt für Geologie und Rohstoffe) (Hrsg.) (2014) Nutzungsdifferenzierte Bodenübersichtskarte 1 : 1 000 000 (BÜK1000N) für Deutschland (Wald, Grünland, Acker). [Use-differentiated soil overview map 1: 1 000 000 (BÜK1000N) for Germany (forest, grassland, arable land)]. https://www.bgr.bund.de/DE/Themen/Boden/Informationsgrundlagen/Bodenkundliche_Karten_Datenbanken/BUEK1000/Nutz_BUEK/nutz_buek_node.html

Bonten LTC, Reinds GJ, Posch M (2016) A model to calculate effects of atmospheric deposition on soil acidification, eutrophication and carbon sequestration. Environ Model Softw 79:75–84

Bösch B (2001) Neue Bonitierungs- und Zuwachshilfen. Schriftenreihe Freiburger Forstliche Forschung, Wissenstransfer in Praxis und Gesellschaft, FVA-Forschungstage 18. https://www.waldwissen.net/technik/inventur/fva_schaetzhilfen/fva_schaetzhilfen.pdf

De Vries W, Kros J, Reinds GJ, Wamelink W, Mol J, van Dobben H, Bobbink R, Emmett B, Smart S, Evans C, Schlutow A, Kraft P, Belyazid S, Sverdrup HU, van Hinsberg A, Posch M, Hettelingh J-P (2007) Developments in deriving critical limits and modelling critical loads of nitrogen for terrestrial ecosystems in Europe, vol 1382. Alterra Green World Research, Report, Wageningen, 206 pp

DWD (Deutscher Wetterdienst) (2012) Mittlere monatliche Niederschlagsmengen und Mittlere Tagesmitteltemperatur der Referenzperiode 1981–2010 für Sommer und Winter. Rasterdatei. [Mean monthly precipitation and mean daily temperature of the reference period 1981–2010 for summer and winter. Raster file]

DWD (Deutscher Wetterdienst) (2021) Mittlere Tagesmitteltemperatur der Referenzperiode (Rasterdatei); Mittlere Niederschlagsmengen aus dem Zeitraum 1991–2020. Rasterdatei [Mean daily temperature of the reference period (raster file); Mean precipitation amounts from the period 1991–2020. Raster file]. https://opendata.dwd.de/climate_environment/CDC/grids_germany/multi_annual/temperature/. Assessed on 23.02.2023

Ellenberg H (1996) Vegetation Mitteleuropas mit den Alpen in ökologischer, dynamischer und historischer Sicht, [Vegetation of Central Europe with the Alps in ecological, dynamic and historical perspective], 5th edn. Ulmer, Stuttgart. 1096 pp

EMEP (2018) Co-operative programme for monitoring and evaluation of the long-range transmissions of air pollutants in Europe. http://www.emep.int/, Accessed 20 Apr 2018

Haines-Young R; Potschin MB (2018) Common International Classification of Ecosystem Services (CICES) V5.1 and guidance on the application of the revised structure. Available from www.cices.eu

Härdtle W (1984) Vegetationskundliche Untersuchungen in Salzwiesen der Ostholsteinischen Ostseeküste. Mitteilungen der AG Geobotanik in Schlesw-Holst. und Hamburg, Kiel, Heft 48, 415 S. [Vegetation studies in salt marshes at the East Holstein Baltic Sea coast].

Härdtle W (1989) Potentiell Natürliche Vegetation - Ein Beitrag zur Kartierungsmethode am Beispiel der Topographischen Karte 1623 Owschlag. Mitteilungen der AG Geobotanik in Schlesw-Holst. und Hamburg, Kiel. Heft 40:73 S. [Potentially natural vegetation – a contribution to mapping methods using the example of Topographic Map 1623 Owschlag.] Mitteilungen der AG Geobotanik in Schlesw-Holst. und Hamburg, Kiel, Vol. 40, 73 pp

Härdtle W (1995a) Vegetation und Standort der Laubwaldgesellschaften (Querco-Fagetea) im Nördlichen Schleswig-Holstein. [Vegetation and location of deciduous forest communities (Querco-Fagetea) in northern Schleswig-Holstein.]. In: Mitteilungen der AG Geobotanik in Schlesw-Holst, vol 48, und Hamburg, Kiel. 415 pp

Härdtle W (1995b) Zur Systematik und Synökologie artenarmer Buchenwälder (Flatter-Gras-/Sauerklee-Buchenwälder) in Schleswig-Holstein. Tuexenia 15:45–51. Göttingen. [On the systematics and synecology of species-poor beech forests (Milio-/Oxalido-Fagetum) in Schleswig-Holstein.] Tuexenia 15, pp. 45–51, Göttingen

Härdtle W, Ewald J, Hölzel N (2004) Wälder des Tieflandes und der Mittelgebirge. Eugen Ulmer, Stuttgart. 250 S. [Forests of the lowlands and low mountain ranges]. Publisher Eugen Ulmer Stuttgart. 250 pp

Hartmann FK, Jahn G (1967) Waldgesellschaften des mitteleuropäischen Gebirgsraumes nördlich der Alpen [Forest communities of the Central European mountain region north of the Alps]. Gustav Fischer, Jena. 636 pp

Hofmann G (1969) Zur pflanzensoziologischen Gliederung der Kiefernforsten des nordostdeutschen Tieflandes [On the phytosociological structure of the pine forests of the north-east German lowlands]. Feddes Repertorium 80:4–6. Berlin, pp 401–412

Hofmann G, Pommer U (2013) Die Waldvegetation Nordostdeutschlands. [The forest vegetation of north-eastern Germany], vol 54. Eberswalder Forstliche Schriftenreihe, Eberswalde, 598 p

Holmberg M., Aherne J., Austnes K., Beloica J., De Marco A., Dirnböck T., Fornasier M.F., Goergen K., Futter M., Lindroos A-J., Krám P., ., Neirynck J., Nieminen T.-M., ., Pecka T., Posch M., Pröll G., Rowe E. C., Scheuschner T, Schlutow A., Valinia S. (2018): Modelling study of soil C, N and pH response to air pollution and climate change using European LTER site observations May 2018 Sci Total Environ 640–641(02):387–399. DOI: https://doi.org/10.1016/j.scitotenv.2018.05.299

Hrivnák R (2002) Aquatic plant communities in the catchment area of the Ipeľ river in Slovakia and Hungary. Part I. Classes Lemnetea and Charetea fragilis. Thaiszia – J. Bot 12:25–50. Online im Internet: http://www.upjs.sk/bz/thaiszia/index.html

Hundt R (1964) Die Bergwiesen des Harzes, Thüringer Waldes und Erzgebirges. Gustav Fischer Publisher, Jena. 284 S. [The Mountain Meadows of the Harz, Thuringian Forest and Ore Mountains]. Gustav Fischer Publisher Jena, 284 pp

Issler E (1924) Les associations végétales des Vosges méridionales et de la plaine rhénane avoisinante. 1. Les forêts (fin). Bull. Soc. Hist. Nat Colmar, Tom 19:1–109

Issler E (1926) Les associations végétales de la partie supérieure de la vallée da la Lane. Thèse, 120 S., Besançon

Issler E (1942) Vegetationskunde der Vogesen. [Vegetation science of the Vosges.] Pflanzensoziologie Band 5, Gustav Fischer Publisher Jena, 161 p

Jacobsen C, Rademacher P, Meesenburg H, Meiwes KJ (2002) Element-Gehalte in Baum-Kompartimenten: Literatur-Studie und Datensammlung [Element contents in tree compartments: Literature study and data collection]. In: Niedersächsische Forstliche Versuchsanstalt, Report. Self-publishing by Forschungszentrum Waldökosysteme der Universität Göttingen, pp 1–80

Klapp E (1954) Wiesen und Weiden. 2., völlig neu gestaltete Aufl., [Meadows and pastures.] 2nd, completely redesigned edition, Publisher Paul Parey, Berlin und Hamburg. 519 p

Klapp E (1965) Grünlandvegetation und Standort—nach Beispielen aus West-, Mittel- und Süddeutschland. [Grassland vegetation and location—according to examples from western, Central and southern Germany.]. Publisher Paul Parey, Berlin/Hamburg. 384 pp

Kranenburg R. (n.d.): PINETI4: Modellierung atmosphärischer Stoffeinträge von 2000 bis 2019 in Deutschland. [PINETI4: Modeling atmospheric pollutant inputs from 2000 to

2019 in Germany] https://www.umweltbundesamt.de/daten/flaeche-boden-land-oekosysteme/land-oekosysteme/ueberschreitung-der-belastungsgrenzen-fuer-0#situation-in-deutschland. Accessed on 19 Dec 2023

Krausch (1962) Der Sandnelken-Kiefernwald an seiner Westgrenze in Brandenburg. Mitteilungen der Floristisch-soziologischen Arbeitsgemeinschaft Neue Folge 9:141–144. [The sand-lark pine forest on its western border in Brandenburg. Mitteilungen der Floristisch-soziologischen Arbeitsgemeinschaft Neue Folge 9: 141–144]

Krieger H (1937) Die flechtenreichen Pflanzengesellschaften der Mark Brandenburg. [The lichen-rich plant communities of Mark Brandenburg]. Beih Bot Centrbl 57:1–76

KWF (Kompetenzzentrum Wald und Forstwirtschaft des Staatsbetriebes Sachsenforst) (2022) Kartierung der Standortsformengruppen der Forstlichen Standortskartierung in Sachsen. [(Competence Centre Forest and Forestry of the State Enterprise Saxony Forest) (2022): Mapping of the site form groups of the forest site mapping in Saxony. Digital dataset. Unpublished]

LfULG (Landesamt für Umwelt, Landwirtschaft und Geologie) (ed) (2020): Regionale Klimaprojektionen für Sachsen. Schriftenreihe des LfULG, Heft 3/2020. [(State Office for the Environment, Agriculture and Geology) (Ed) (2020): Regional climate projections for Saxony. LfULG publication series, issue 3/2020] https://publikationen.sachsen.de/bdb/artikel/35082

Liebert H-P (1988) Umwelteinfluss auf Wachstum und Entwicklung von Wasserpflanzen sowie deren Rolle bei der Reinhaltung unserer Gewässer. [Environmental influence on the growth and development of aquatic plants and their role in keeping our waters clean]. Bibliografische Mitteilungen der Universität Jena 35, Jena

Lohmeyer W (1957) Der Hainmieren-Schwarzerlenwald [Stellario-Alnetum glutinosae (Kästner 1938)]. [The grove alder forest [Stellario-Alnetum glutinosae (Kästner 1938)]]. Mitt. flor.-soz. Arb.gem. N. F 6/7:247–257. Stolzenau/W

Lohmeyer W (1962) Zur Gliederung der Zwiebelzahnwurz (Cardamine bulbifera)-Buchenwälder im nördl. Rheinischen Schiefergebirge. Mitteilungen der Floristisch-soziologischen Arbeitsgemeinschaft Neue Folge 9:187–193

LWF (Bayrische Landesanstalt für Wald und Forstwirtschaft) (2024) Baumartenporträts der wichtigsten Waldbäume in Bayern [Bavarian State Institute for Forestry and Forest Economics) (2024): Tree species portraits of the most important forest trees in Bavaria]. https://www.lwf.bayern.de/waldbau-bergwald/waldbau/094341/index.php

Madl P (2017) Vorlesungsskript Bodenökologie, gelesen von W. Strobl, Universität Salzburg [Lecture notes on soil ecology, read by W. Strobl, University of Salzburg]. http://biophysics.sbg.ac.at/transcript/boden.pdf

Mahn E-G (1959) Vegetations- und standortskundliche Untersuchungen an Felsfluren, Trocken- und Halbrocken rasen Mitteldeutschlands. [Vegetation and site studies of rocky meadows, dry and semi-arid grasslands in Central Germany] Diss. Uni Halle, 215 p

Mahn E-G (1965) Vegetationsaufbau und Standortsverhältnisse der kontinental beeinflussten Xerothermrasengesellschaften Mitteldeutschlands [Vegetation structure and site conditions of the continental-influenced xerothermic grass communities of Central Germany]. In: Abhandlungen der Sächsischen Akademie der Wissenschaften zu Leipzig. Akademie-Publisher, Berlin, 138 p

Matuszkiewicz H (1962) Zur Systematik der natürlichen Kiefernwälder des mittel- und osteuropäischen Flachlandes. Mitteilungen der Floristisch-soziologischen Arbeitsgemeinschaft Neue Folge 9:145–186

Matuszkiewicz H, Borowik A (1958) Zur Systematik der Auenwälder in Polen. Acta Soc Bot Pol XXVI-Nr. 4:719–756

Matuszkiewicz H, Matuszkiewicz N (1956) Pflanzensoziologische Untersuchungen im Forstrevier „Ruda" bei Pulawy (Polen). Acta Soc Bot Pol XXV-Nr. 2: 331–400

Matuszkiewicz H, Traczyk T (1958) Zur Systematik der Bruchwaldgesellschaften (Alnetalia glutinosae) in Polen. Acta Soc Bot Pol XXVII-Nr. 1:21–44

Nagel H-D, Schlutow A, Kraft P, Scheuschner T, Weigelt-Kirchner R (2010) Modellierung und Kartierung räumlich differenzierter Wirkungen von Stickstoffeinträgen in Ökosysteme im Rahmen der UNECE-Luftreinhaltekonvention. Teilbericht II Das BERN-Modell— ein Bewertungsmodell für die oberirdische Biodiversität. [Modelling and Mapping of Spatially Differentiated Effects of Nitrogen Inputs to Ecosystems under the UNECE Clean Air Convention Sub-report II The BERN model—an assessment model for above-ground biodiversity]   UBA-Texte   08/2010.   https://www.umweltbundesamtde/publikationen/modellierung-kartierung-raeumlich-differenzierter-0

Oberdorfer E (1957) Süddeutsche Vegetationsgesellschaften [South German plant communities]. Pflanzensoziologie 10, 564 S. Jena

Oberdorfer E (Hrsg.) (1992–1998): Süddeutsche Vegetationsgesellschaften. Teil I 4. Aufl. 1998, Teil II 3. Aufl. 1993, Teil III 3. Aufl. 1993, Teil IV 1992, Gustav-Fischer-Publisher, Jena Stuttgart New York, Teile I-IV in 5 Bänden. [South German Vegetation Communities. Part I 4th ed. 1998, Part II 3rd ed. 1993, Part III 3rd ed. 1993, Part IV 1992, Gustav-Fischer-Publisher, Jena Stuttgart New York, Parts I-IV in 5 volumes]

Oberdorfer E (2001) Pflanzensoziologische Exkursionsflora für Deutschland und angrenzende Gebiete, 8. Aufage. [Sociology of Plants for Excursions in Germany and Adjacent Areas, 8th ed.], Ulmer Publisher, Stuttgart, 1051 S

Orlowsky B, Gerstengarbe FW, Werner PC (2008) A resampling scheme for regional climate simulations and its performance compared to a dynamical RCM. Theor Appl Climatol 92:209–223

Passarge H (1960) Waldgesellschaften NW-Mecklenburgs. [Forest communities of North-West Mecklenburg]. Arch Forstwesen (Berlin) 9:499–541

Passarge H (1964) Vegetationsgesellschaften des nordostdeutschen Flachlandes I. [Vegetation communities of the northeast German lowlands I.]. Pflanzensoziologie 13, Jena. 324 p

Passarge H, Hofmann G (1968) Vegetationsgesellschaften des nordostdeutschen Flachlandes. II. . [Vegetation communities of the northeast German lowlands II]. Pflanzensoziologie 16, Jena. 298 pp. + Annex

Pott R (1992) Die Pflanzengesellschaften Deutschlands, [The plant communities of Germany]. Ulmer, Stuttgart. 427 p

Pottgiesser T, Sommerhäuser M (2004) Fließgewässertypologie Deutschlands: Die Gewässertypen und ihre Steckbriefe als Beitrag zur Umsetzung der EU-Wasserrahmenrichtlinie. [Germany's stream typology: The water body types and their profiles as a contribution to the implementation of the EU Water Framework Directive]. In: Steinberg C, Calmano W, Wilken R-D, Klapper H (eds) Handbook of limnology. 19th Delivery. 7/04. VIII-2.1: 1–16 + Annex

Preising E (1953) Süddeutsche Borstgras- u. Zwergstrauchheiden (Nardo-Callunetea). [South German bristly grass and dwarf shrub heaths (Nardo-Callunetea)]. Mitt. flor. soz. Arbeitsgem.N.F. 4: 112-123. Stolzenau

Preising E, Vahle H-C, Brandes H, Hofmeister H, Tüxen J, Weber HE (1990a) Die Pflanzengesellschaften   Niedersachsens—Bestandsentwicklung,   Gefährdung   und Schutzprobleme: Salzpflanzengesellschaften der Meeresküsten und des Binnenlandes. [The plant communities of Lower Saxony—population development, endangerment and conservation problems: Salt plant communities of the sea coasts and inland areas] Naturschutz und Landschaftspflege Niedersachsens, Heft 20/7 (1–161), Hannover

Preising E, Vahle H-C, Brandes H, Hofmeister H, Tüxen J, Webe, HE (1990b) Die Pflanzengesellschaften   Niedersachsens—Bestandsentwicklung,   Gefährdung   und Schutzprobleme: Wasser- und Sumpfpflanzengesellschaften des Süßwassers. [The plant communities of Lower Saxony—population development, endangerment and conservation problems: freshwater aquatic and marsh plant communities] Naturschutz und Landschaftspflege Niedersachsens, Heft 20/8 (1–161), Hannover

Preising E, Vahle H-C, Brandes H, Hofmeister H, Tüxen J, Weber HE (1997) Die Pflanzengesellschaften   Niedersachsens—Bestandsentwicklung,   Gefährdung   und Schutzprobleme: Rasen-, Fels- und Geröllgesellschaften. [The plant communities of Lower Saxony – population development, endangerment and conservation problems: Grass, rock and boulder communities] Naturschutz und Landschaftspflege Niedersachsens, Heft 20/5 (1–146), Hannover

Roberts DW (1986) Ordination on the basis of fuzzy set theory. Vegetatio 66:123–131

Schachtschabel P, Auerswald K, Brümmer G, Hartke KH, Schwertmann U (1998) Lehrbuch der Bodenkunde [Textbook of soil science]. Verlag Ferdinand Enke, Stuttgart

Scheffer F, Ulrich B (1960) Humus und Humusdüngung [Humus and humus fertilization]. Zweite, völlig neu bearbeitete Auflage [Second, completely reworked edition]. Band I: Morphologie, Biologie, Chemie und Dynamik des Humus. Mit 45 Abbildungen und 39 Tabellen. 1960. VII, 266 Seiten [Volume I: Morphology, Biology, Chemistry and Dynamics of Humus. With 45 figures and 39 tables. 1960. VII, 266 p.]

Schlutow A (2007) The BERN model. In: De Vries W, Kros H, Reinds GJ, van Dobben H, Hinsberg A, Schlutow A, Sverdrup H, Butterbach-Bahl K, Posch M, Hettelingh JP (eds) Developments in deriving critical limits and modelling critical loads of nitrogen for terrestrial ecosystems in Europe, Alterra Report 1382 [Alterra-rapport 1382]. Wageningen University, Wageningen. Alterra pp 64–73 and Annex 4

Schlutow A, Gemballa R (n.d.) Ableitung von klimawandelangepassten Leitwaldgesellschaften für die Wälder im Freistaat Sachsen [Derivation of climate change-adapted indicator forest communities for the forests in the Free State of Saxony]. Staatsbetrieb Sachsenforst (ed.). Pirna. in Preparation

Schlutow A, Hübener P (2002) The BERN model: bioindication for ecosystem regeneration within natural conditions In: Achermann B, Bobbink R (eds) (2003) Empirical critical loads for Nitrogen. Proceedings of the Expert workshop in Berne 11–13 November 2002. Environmental Documentation No 164–Air (Swiss Agency for the Environment, Forests and Landscape SAEFL) p 305-311

Schlutow A, Hübener P (2004) The BERN Model bioindication for ecosystem regeneration towards natural conditions UBA-Texte 22/04 Umweltbundesamt Berlin, 50 p

Schlutow A, Scheuschner T (2023) Determination of critical loads for eutrophying and acidifying air pollutant inputs for the protection of near-natural ecosystems in Germany. Atmos 14(2):383. https://doi.org/10.3390/atmos14020383

Schlutow A, Schröder W (2021) Rule-based classification and mapping of ecosystem services with data on the integrity of forest ecosystems. Environ Sci Eur 33:50. https://doi.org/10.1186/s12302-021-00481-3

Schlutow A, Dirnböck T, Pecka T, Scheuschner T (2015) Chapter 14: Use of an empirical model approach for modelling trends of ecological sustainability. In: De Vries W, Hettelingh J-P, Posch M (eds) Critical loads and dynamic risk assessments: Nitrogen, acidity and metals in terrestrial and aquatic ecosystems. Springer. 662 S

Schlutow A, Bouwer Y, Nagel H-D (2018) Bereitstellung der Critical Load Daten für den Call for Data 2015–2017 des Coordination Centre for Effects im Rahmen der Berichtspflichten Deutschlands für die Konvention über weitreichende grenzüberschreitende Luftverunreinigungen (CLRTAP). Im Auftrag des UBA, Abschlussbericht Projekt-Nr. UBA/43848. [Provision of critical load data for the Call for Data 2015–2017 of the Coordination Centre for Effects in the context of Germany's reporting obligations for the Convention on Long-Range Transboundary Air Pollution (CLRTAP). On behalf of UBA, Final Report Project no. UBA/43848]. https://www.umweltbundesamt.de/publikationen/critical-load-daten-fuer-die-berichterstattung-2015

Schlutow A, Schröder W, Scheuschner T (2021) Assessing the relevance of atmospheric heavy metal deposition with regard to ecosystem integrity and human health in Germany. Environ Sci Eur 33(1):1–34. https://doi.org/10.1186/s12302-020-00391-w

Schlutow A, Kraft P, Scheuschner T, Schlutow M, Schröder W (2024) Bioindication for Ecosystem Regeneration towards Natural conditions—the BERN data base and BERN model. Environ Sci Eur 36(1):1–21. https://doi.org/10.1186/s12302-023-00826-0. BERN database (open source) under. https://github.com/bern-model/BERN

Schmidt PA, Hempel W, Denner M, Döring N, Gnüchtel B, Walter B, Wendel D (2002) Potentielle natürliche Vegetation Sachsens mit Karte 1:200.000. [Potential Natural Vegetation of Saxony with Map 1:200.000]. Saxon State Office for Environment and Geology, Dresden, 230 p

Schröder W, Schlutow A, Dworcyk C, Jenssen M, Nickel, S (2020) Regelbasierte Einstufung und Kartierung von Ökosystemleistungen mit Daten zur Integrität von Waldökosystemtypen. In book: Handbuch der Umweltwissenschaften. 28. Erg. Lfg. [Rule-based classification and mapping of ecosystem services with data on the integrity of forest ecosystem types. In book: Handbook of Environmental Sciences]

Schröder W, Nickel S, Dreyer A, Völksen B (2023) Accumulation of atmospheric metals and nitrogen deposition in mosses: temporal development between 1990 and 2020, Comparison with emission data and tree canopy drip effects. Pollutants 3:89–101

Schubert R (1960) Die zwergstrauchreichen azidiphilen Pflanzengesellschaften Mitteldeutschlands, [The dwarf-shrub-rich acidophilous plant communities of Central Germany]. VEB Gustav Fischer, Jena. 235 p

Schubert R (1991) Lehrbuch der Ökologie [Textbook of ecology]. Publisher Fischer, Jena. 657 p

Schubert R, Klotz W, Hilbig S (1995) Bestimmungsbuch der Pflanzengesellschaften Mittel- und Nordostdeutschlands. [Identification book of the plant communities of Central and North-Eastern Germany]. Fischer, Jena. 403 pp

Schulte-Bisping H, Beese F (2016) N-Fluxes and N-turnover in a mixed beech-pine forest under low N-inputs. European Journal of Forest Research, vol 135. Springer, Berlin/Heidelberg, pp 229–241. https://doi.org/10.1007/s10342-015-0931-x

Schulze G (1998) Anleitung für die forstliche Standortserkundung im nordostdeutschen Tiefland (St andortserkundungsanleitung) - SEA 95—Teil D. - Bodenformen-Katalog Merkmalsübersichten und –tabellen für Haupt- und Feinbodenformen [Guidance for forest site reconnaissance in the north-east German lowlands—SEA95—Part D. Soil Shapes Catalogue Feature Overviews and tables for main and fine soil forms], 4th edn, Schwerin

Slobodda S (1982) Pflanzengesellschaften als Kriterium zur ökologischen Kennzeichnung des Standortsmosaiks [Plant communities as a criterion for the ecological characterisation of the site mosaic]. Archiv Naturschutz und Landschaftspflege 22(2):79–101

SMUL (Staatsministerium für Umwelt und Landwirtschaft des Freistaates Sachsen) (2013) Waldstrategie 2050. [Forest strategy 2050] https://www.publikationen.sachsen.de ›bdb›documents›Waldstrategie2050-Publikationen-sachsen.de

Succow M (1974) Vorschlag einer systematischen Neugliederung der mineralbodenwasserbeeinflussten wachsenden Moorvegetation Mitteleuropas unter Ausklammerung des Gebirgsraumes [Proposal of a systematic reclassification of the mineral soil water-influenced growing mire vegetation of Central Europe excluding the mountain area]. In: Feddes Repertorium, Vol. 85, Heft 1-2:57-113. Berlin

Succow M (1988) Landschaftsökologische Moorkunde [Landscape ecology of peatlands]. Gustav-Fischer-Publisher, Jena. 126 pp

Succow M, Joosten H (2001) Landschaftsökologische Moorkunde [Landscape ecology of peatlands], 2nd edn. Schweizerbart'sche Publishersbuchhandlung, Stuttgart. 622 pp

Tüxen R (1937) Die Pflanzengesellschaften Nordwestdeutschlands [The plant communities of Northwest Germany]. Mitt. flor.-soz. Arb.gem. Niedersachsen 3:1-170. Hannover

Tüxen R (1958) Pflanzengesellschaften oligotropher Heidetümpel Nordwestdeutschlands [Plant communities of oligotrophic heathland ponds in Northwest Germany]. Publishing by Geobotanisches Institut Rübel in Zürich 33. Bern

Tüxen R, Westhoff V (1963) Saginetea maritimae, eine Gesellschaftsgruppe im wechselhalinen Grenzbereich der europäischen Meeresküsten [Saginetea maritimae, a social group in the alternate haline boundary region of the European seacoasts]. In: Mitt. flor.-soz. Arbgemeinsch. 1963, N. F. 10:116-129. Stolzenau/Weser

UBA—Umweltbundesamt (2015) Corine Land Cover - Bodenbedeckungsdaten für Deutschland [Land cover data for Germany] CORINE 2012, hochaufgelöste Version [high resolution version] LBM-DE2012 © BKG/Geobasis-DE

Volk OH (1937) Über einige Trockenrasengesellschaften des Würzburger Wellenkalkgebietes [About some dry grassland communities of the Würzburg Wavy Limestone area]. Beih. Bot. Cbl., 57 (1937), pp. 577-598 Abt. B

von Rochow M (1951) Die Pflanzengesellschaften des Kaiserstuhls. [The plant communities of the Kaiserstuhl]. Pflanzensoziol 8. 140 pp. Jena

Walz U, Stein C (2014) Indicators of hemeroby for the monitoring of landscapes in Germany. J Nat Conserv 22(3):279–289. https://www.researchgate.net/publication/265692282_Die_Naturlichkeit_der_Landnutzung_in_Deutschland?channel=doi&linkId=54194c020cf25eb ee9884369&showFulltext=true

Wellbrock N, Bolte A, Flessa H (Hrsg.) (2017) Dynamik und räumliche Muster forstlicher Standorte in Deutschland - Ergebnisse der Bodenzustandserhebung im Wald. [Dynamics and spatial patterns of forest sites in Germany - results of the soil condition survey in forests]. https://www.thuenen.de/media/institute/wo/Waldmonitoring/bze

Willner W (2002) Syntaxonomische Revision der südmitteleuropäischen Buchenwälder [Syntaxonomic Revision of the South-Central European Beech Forests]. Phytocoenologia 32(3) Berlin-Stuttgart, p. 337–453

Willner W, Grabherr G (eds) (2007) Die Wälder und Gebüsche Österreichs [The Forests and Shrublands of Austria]. Elsevier, Spektrum Akademic Publisher, Heidelberg. München., Textband 302 S., Tabellenband 290 S

Wolfram C (1996) Die Vegetation des Bottsandes [The vegetation of the Bottsand]. Mitt. der AG Geobotanik in Schlesw-Holst. und Hamburg, Kiel, Heft 51, 111 pp

Chapter 4
Discussion

Abstract For a validation of the rule-based rating of the three ecosystem services examined in more depth, the classification and mapping of 135 current near-natural ecosystem types (ANOEST) in Germany derived from vegetation and soil data of about 22,000 forest stands (Schröder W, Schlutow A, Dworcyk C, Jenssen M, Nickel S, Regelbasierte Einstufung und Kartierung von Ökosystemleistungen mit Daten zur Integrität von Waldökosystemtypen. In Handbuch der Umweltwissenschaften. 28. Erg. Lfg. [Rule-based classification and mapping of ecosystem services with data on the integrity of forest ecosystem types. In: Handbook of environmental sciences, 2020; Schlutow A, Schröder W, Jenssen M, Nickel S, Environ Sci Eur 33:87, 2021), was compared with the results at hand.

Keywords ANOEST (current near-natural ecosystem types) · Rule-based rating

4.1 Comparison of Habitat Availability Assessments Using the Rule-Based Approach (Sect. 2.2) with the Corresponding Information in the ANOEST Ecosystem Type Dataset

The results of the rule-based rating of the habitat service scores of Germany's ecosystem type classes (Schlutow and Schröder 2021) are compared with the rating scores of the current near-natural ecosystem types (ANOEST) for Germany in Schröder et al. (2020).

The ANOEST attribute dataset of ecosystem types (Schröder et al. 2020) contains a verbal-argumentative assessment of habitat service (moderate, medium, high, very high) for each of the 135 current near-natural ecosystem types (ANOEST). The categorisation was based on expert judgments. The closeness to nature of the vegetation was indicated using the Kullback distance, but only for 60 of the 135 ANOEST.

In order to carry out a statistical comparison of the ANOEST expert judgments with the rule-based assessment results, the verbal assessments for the ANOEST dataset were converted to the ordinal scale of 0–5 by making the following assignment: very high = 5, high = 4, medium = 3, moderate = 2. However, the assessment "medium" does not occur in the ANOEST assessment results. The graphical representation of the comparison can be found in Fig. 4.1.

The Spearman correlation coefficient (linear correlation) is 0.86. The regression line does not deviate significantly from the 1:1 line (t(paired) = 125). According to Brosius et al. (2016), this results in a very strong agreement between the two methods. Only in 4 ANOEST are there clear differences in the evaluation. The sycamore-ash forest is assessed 3 points higher than the assessment according to the rule-based approach (Table 3.12). The reasons for the low score are not clear. It is an FFH habitat type with very high naturalness, relative species diversity of the flora and habitat suitability for animals. The brown gauze oak forest, the wet meadow oak forest and the wet meadow oak forest are each rated 2 points higher according to the rule-based approach (4 vs. 2 points) (Table 3.12). The lower score is due to the fact that these are forests. Pedunculate and/or sessile oaks are site-appropriate and native, but are overrepresented in a forest. The closeness to nature would therefore only be medium (3), but some of the other criteria could be rated as good (4 points).

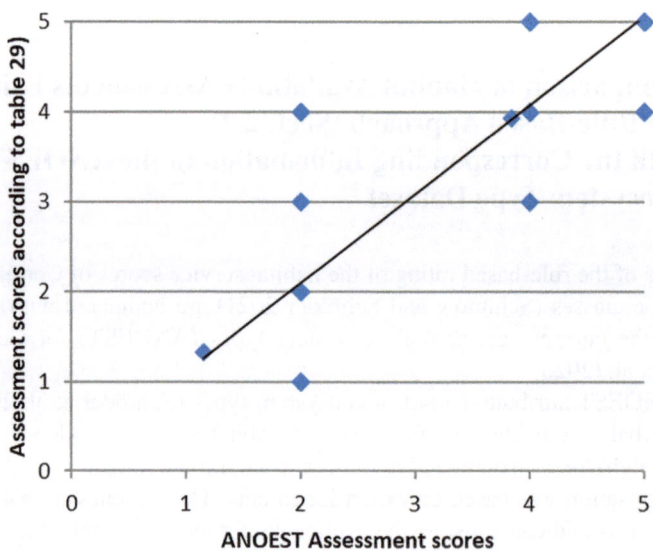

Fig. 4.1 Comparison of the assessment results of the habitat service according to Table 3.12 with the ANOEST assessment results. n = 125 paired, many points on top of each other. (Source: author's own illustration)

4.2 Comparison of the Estimates of Potential Biomass Primary Productivity from the Approach in Sect. 2.3 with the ANOEST Ecosystem Type Dataset

The ANOEST attribute dataset for the current ecosystem types contains information on the annual growth rate of tree wood at the time of the peak (jZR) for 60 ANOEST. The information on net primary production (NPP) in the ANOEST description is based on the statistical analysis of a large number of empirically collected data. In order to be able to carry out a statistical comparison of the two evaluation methods, the data on net primary biomass production (NPP) was converted into value levels from 0 to 5, whereby it is assumed that the smallest value cannot be 0, but is evaluated with 1 point. Between the NPP minimum (= 1 point) and the NPP maximum (= 5 points) of all 60 ANOEST values, the evaluation score was calculated by interpolation with the ANOEST-specific NPP according to the following formula:

$$NPP = \left(jZR\left(ANOEST\right) - jZR\left(min\right)\right) / \left(jZR\left(max\right) - jZR\left(min\right)\right) * 4 + 1$$

With:

NPP = biomass primary productivity Assessment points
The graphical representation of the comparison can be found in Fig. 4.2.

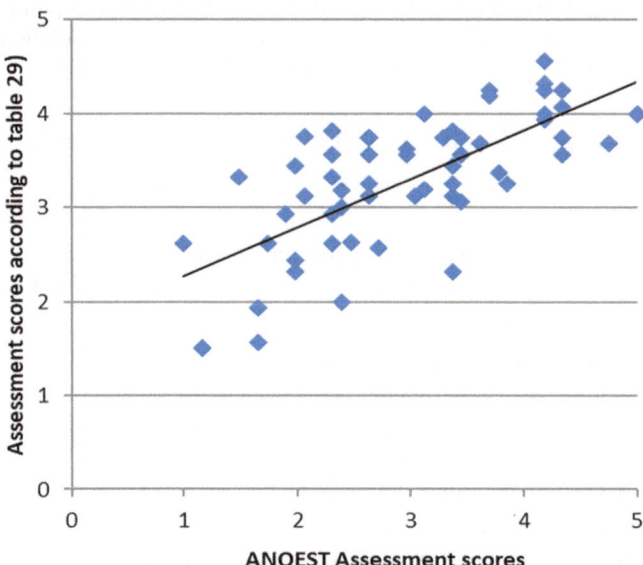

Fig. 4.2 Comparison of the assessment points of the potential biomass primary productivity according to Table 3.12 with the ANOEST assessment points, n = 60 paired. (Source: author's own presentation)

The Spearman correlation coefficient (linear correlation) is 0.72. According to Brosius et al. (2016), this results in a strong agreement of the evaluation according to Table 3.12 with the measured values of the growth rate in the ANOEST database. The regression line deviates slightly, but not significantly, from the 1:1 line (t(paired) = −5.0). The difference in the standard deviations of the two data series is very small (0.7 points). Although there are differences in the data basis for the plant physiological yield potential (annual growth rate at the time of culmination compared to the annual growth rate averaged over 100 years), there is still a high level of agreement.

4.3   Comparison of the Assessments of Potential Carbon Sequestration from the Rule-Based Approach According to Sect. 2.4 with the ANOEST Ecosystem Type Dataset

The ANOEST attribute dataset contains information on the C_{org} content in the soil profile up to a depth of 80 cm for 60 ecosystem types. The information on organic carbon stocks in the ANOEST description is based on the statistical analysis of a large amount of empirically collected data.

In order to be able to carry out a statistical comparison of the two evaluation methods, the information on the C content$_{org}$ was converted into value levels from 0 to 5. It is assumed that the lowest value cannot be 0, but is assessed with 1 point. Between the C_{org} minimum (=1 point) and the C_{org} maximum (=5 points) of all 60 ANOEST values, the evaluation score is calculated by interpolation with the ANOEST-specific C_{org} content according to the following formula:

$$K = \left(C_{org}\left(ANOEST\right) - C_{org}\left(\min\right)\right) / \left(C_{org}\left(\max\right) - C_{org}\left(\min\right)\right) * 4 + 1$$

With:

K = Evaluation points for the carbon sequestration service
The graphical representation of the comparison can be found in Fig. 4.3.

The Spearman correlation coefficient (linear correlation) is 0.54. According to Brosius et al. (2016), this results in an average agreement of the assessment according to both methods. However, the linear regression line deviates significantly from the 1:1 line (t(paired) = −10.5). It rather describes the left half of a parabola. The difference between the standard deviations of the two data series is 1.15 points.

The differences are particularly clear in the low scores for ANOEST, which were more strongly differentiated according to the rule-based approach (Sect. 2.4) and in some cases were rated up to three points higher. This is due to the high C_{org} maximum values from the data sheets, with only three ANOEST C_{org} values above the mean. While the gauze carbonate soils are rated highly using the rule-based approach, they only have average values according to the ANOEST database. Only

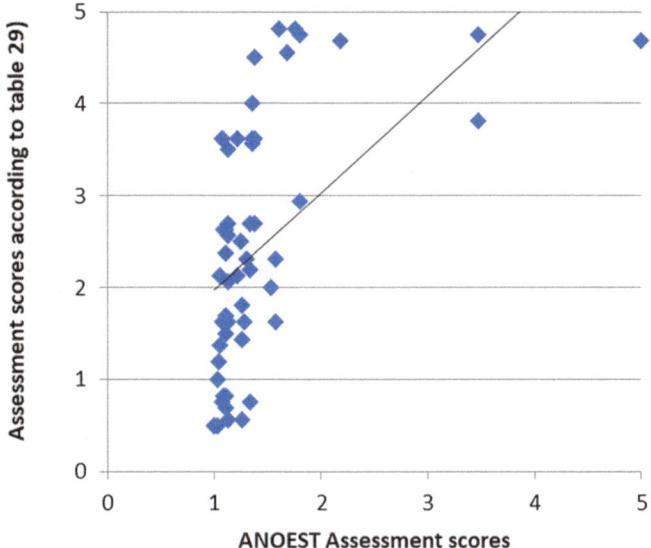

Fig. 4.3 Comparison of the assessment results of potential carbon sequestration according to Table 3.12 with the ANOEST assessment results, n = 125 paired. (Source: Author's own illustration)

three black alder wet bogs ANOEST are rated with high value points, which were also rated very high according to the rule-based approach. In contrast, a number of other moist and wet ANOESTs have only low C_{org} contents. The ANOEST attribute datasets therefore do not indicate a general trend of higher C sequestration capacity with higher soil water content.

It is of course difficult to explain carbon sequestration with just a few influencing factors. However, it is also possible that only one rule for combining the individual criteria assessments is equally inadequate for all ANOEST. In the evaluation according to the rule-based approach, the C sequestration service in the mineral soil was prioritised because the proportion of C_{org} in the mineral soil of the second forest soil inventory sites (Wellbrock et al. 2017) was three times higher than in the humus. This could overestimate the C sequestration capacity in the mineral soils compared to some organic wet soils. However, there is no generalisable evidence for an adjustment of the rule from the reference values of the C_{org} content.

References

Brosius H-B, Haas A, Koschel F (2016) Methoden der empirischen Kommunikationsforschung. Eine Einführung [Methods of empirical communication research. An introduction]. Edition: 7. überarbeitete und aktualisierte Auflage. Springer VS. https://doi.org/10.1007/978-3-531-19996-2

Schlutow A, Schröder W (2021) Rule-based classification and mapping of ecosystem services with data on the integrity of forest ecosystems. Environ Sci Eur 33:50. https://doi.org/10.1186/s12302-021-00481-3

Schlutow A, Schröder W, Jenssen M, Nickel S (2021) Modelling of soil characteristics as basis for projections of potential future forest ecosystem development under climate change and atmospheric nitrogen deposition. Environ Sci Eur 33:87

Schröder W, Schlutow A, Dworcyk C, Jenssen M, Nickel S (2020) Regelbasierte Einstufung und Kartierung von Ökosystemleistungen mit Daten zur Integrität von Waldökosystemtypen. In: Handbuch der Umweltwissenschaften. 28. Erg. Lfg. [Rule-based classification and mapping of ecosystem services with data on the integrity of forest ecosystem types. In: Handbook of environmental sciences]

Wellbrock N, Bolte A, Flessa H (Hrsg.) (2017) Dynamik und räumliche Muster forstlicher Standorte in Deutschland – Ergebnisse der Bodenzustandserhebung im Wald. [Dynamics and spatial patterns of forest sites in Germany – results of the soil condition survey in forests]. https://www.thuenen.de/media/institute/wo/Waldmonitoring/bze

Chapter 5
Conclusions

Abstract The presented methodology allows a transparent and, thus, reproducible quantitative assessment and mapping of ecosystem services in the pristine condition of wood ecosystems, which can be compared with those under current and expected future conditions considering nitrogen and sulphur deposition and climate change. The ordinal classification of ecosystem service potentials is used to derive the necessity and the type and scope of measures to restore the highest possible performance from the comparison of ecosystem service potentials in the pristine condition with current and expected future ecosystem service potentials. The rule-based classification of ecosystem types proposed enables the transfer of the methodology to areas outside Germany, especially to Central Europe. The methodology is also applicable to other spatial scales.

Keywords Reproducible assessment · Cartographic scales

5.1 Ecosystem Services for Germany Analysed in Detail

The assessment of the habitat service of the ecosystem in its current state (Fig. 3.9, Table 3.11) shows that in Germany's ecosystem type classes with near-natural mixed forests on rare extreme sites, for example, the moist and wet forests and the near-natural forests of nutrient-poor and/or dry sites have the highest ecosystem service values. In addition to the high hemerobia rating, the high protection status also contributes to this. Coniferous afforestations, on the other hand, generally have low scores. However, if they correspond to the potentially natural vegetation in terms of tree species composition and differ from it only in the overrepresentation of the main conifer species and the absence of secondary tree species, they can also achieve medium scores. The occurrence of high to very high habitat potential is distributed throughout Germany and is particularly concentrated in the south-west and in the Alps. The areas with low and very low potential are found in the Harz Mountains, parts of Bavaria and eastern Germany.

© The Author(s), under exclusive license to Springer Nature
Switzerland AG 2024
A. Schlutow, W. Schröder, *Climate Change and Atmospheric Deposition as Drivers of Forest Ecosystem Integrity and Services*, SpringerBriefs in Environmental Science, https://doi.org/10.1007/978-3-031-67103-6_5

The presented databases (Schlutow and Schröder 2021; Schlutow et al. 2024) provide at least a good overview of the potential services of the current near-natural forest ecosystems as habitats for native wild species at a scale of 1:1 million across Germany. The tables categorising the German forest ecosystem types into classes also enable the user to make a corresponding classification for a large-scale area mapping or for individual sites, as the criteria can be checked and the result can thus be repeated.

The results of the assessment of the individual criteria (after Schlutow et al. 2021) and the overall assessment of current primary biomass production (Fig. 3.10) lead to the following conclusion: The ecosystem classes with high and very high potential are "sub-oceanic, moderately dry to fresh, moderately nutrient-rich beech forest", "sub-oceanic, moderately dry to fresh, moderately nutrient-rich spruce forest" and "sub-oceanic, moderately dry to fresh, nutrient-rich beech forest". This potential is mainly concentrated in the low mountain ranges in southern and western Germany. A low to very low potential for primary biomass production was determined for the ecosystem class "Central European to subcontinental, dry, rather poor pine forest". The areas with low potential are concentrated in northern Saxony (Lower Lusatia) and throughout Brandenburg.

The data from the BÜK1000N and the climate data from the DWD appear to be a sufficient basis for estimating potential net primary productivity at a scale of 1:1 million. The map of Germany cannot be used without further ado for larger scales or for a site-specific estimate. Only when a detailed examination has shown that the reference soil profile of the BÜK1000N is actually sufficiently present at the study site can the yield estimate from the map of Germany be transferred to individual sub-areas or sites. If there is an atypical example of a soil type at the study site that is not represented in the mapped soil unit of the BÜK1000N due to the scale, an analogue extrapolation from another area to the study site is possible, or the calculation is carried out using data collected elsewhere using the proposed method.

From the results of the assessment of current carbon sequestration for ecosystem services (Fig. 3.11), it can be deduced that the highest classifications are assigned to rich beech forests such as "central European to subcontinental, moderately dry to fresh, nutrient-rich hornbeam forest", "suboceanic, moderately dry to fresh, nutrient-rich beech forest" and "suboceanic, moist, nutrient-rich beech forest", especially in the Swabian and Franconian Alb and in the Thuringian Basin. Pine forests and pine woodlands, which are mostly restricted to dry or moist, nutrient-poor sites, have a low carbon storage capacity. The ecosystem type classes "Central European to subcontinental, dry, moderately nutrient-rich pine forest" and "Central European to subcontinental, dry, rather nutrient-poor pine forest", which are considered to have a very low potential for carbon storage in the soil, are particularly common. These classes are particularly common in Brandenburg and Saxony. The 25 ecosystem classes with a low carbon storage capacity currently include near-natural forest ecosystems, such as the ecosystem class "suboceanic, moderately dry to fresh, rather poor beech forest". However, many ecosystem classes also have a low potential for carbon sequestration. These ecosystem classes include, for example, spruce forests such as the "suboceanic, moderately dry to fresh, moderately nutrient-rich spruce forest" or the "suboceanic, moderately dry to fresh, moderately nutrient-rich pine forest", which are very dominant in terms of area. In addition, 17 ecosystem classes have a medium potential for carbon sequestration.

The most important risk factor to which forests are exposed is pollution from atmospheric N inputs (Schlutow and Scheuschner 2023). While for centuries the export of biomass through intensive timber extraction, litter utilisation and forest grazing reduced the N stocks of many forest soils and the N supply was the most important growth-limiting site factor for a long time, this situation has changed fundamentally in recent decades due to the atmospheric deposition of nitrogen oxides from industry and traffic as well as ammonium from agriculture (Wellbrock et al. 2017). In most soils, exceeding the critical nitrogen loads is associated with a leaching of basic cations into deeper soil layers and even leaching into the groundwater, meaning that the supply of basic nutrients to trees can now become a limiting factor. However, a large proportion of the forest soils in Germany that are at risk of acidification have been limed, i.e., supplied with the nutrients calcium, magnesium and potassium. As a result, the growth rates of wood of all dominant tree species in Germany almost doubled between 1975 and 2000. The application of the rule-based evaluation system for net biomass production using the wood growth rates surveyed in 2000 (Bösch 2001) shows an increase in evaluation points of at least one point on all forest soils in the lowlands as well as in the medium and low mountain ranges in Germany. Only in the higher mountain regions did the spruce forests that predominate there not experience an increase in assessment points.

The extent of C sequestration in terrestrial ecosystems is affected by increased N inputs when they are N-limited. As long as forest ecosystems have been N-limited in the past, they react sensitively to N inputs, e.g., by increasing plant biomass production and thus the input of plant litter into the soil, which in turn can lead to C sequestration (Wellbrock et al. 2017). An assessment of the current condition of the forests as part of the soil condition surveys (Wellbrock et al. 2017) could only be carried out to a limited extent, meaning that no reliable statements can be made in this regard. However, forest sites with N deficiency are currently hard to find.

With regard to N inputs, Dirnböck et al. (2017) were able to show that the immigration of nitrophytes is favoured when critical load limits are exceeded by N deposition. In the data set of the soil condition survey 2006–2008 (Wellbrock et al. 2017), a correlation between N pollution and composition of the herb layer is only evident for the records of the blueberry spruce forest (suboceanic, moderately dry to fresh, nutrient-poor spruce forest/wood). Here, both the number of nitrophytes and the number of species in the forest edges and clearings increase with the difference between N deposition and N uptake capacity of the forest ecosystems. The effect on montane spruce forests can be explained by the fact that the N input is relatively high due to the high precipitation in the montane region, while the N uptake capacity of the tree population is rather low due to the low temperatures.

The provision of ecosystem services is based on complex interactions between biotic and abiotic ecosystem components, which can be recorded using quantitative data from environmental monitoring and a classification of ecosystem integrity based on this. The comprehensive methodology presented operationalises the specifications of the MAES working group quantitatively (MAES 2016). The MAES classification framework for integrative ecosystem assessments comprises the mapping of ecosystems, the classification of ecosystem states (ecosystem state information for individual indicators, ecosystem functions and ecosystem types), the classification of ecosystem services and their integration. The rule-based classification of the three ecosystem services analysed in depth using quantitative indicators is unique in the EU to date.

5.2   In-Depth Study of Ecosystem Services for the Kellerwald National Park Special Area

As part of the assessment of habitat services, the category "no significant potential" and "medium potential" was not identified in any ecosystem type class within the area. The ecosystem type class "suboceanic, moderately dry to fresh, poor larch forest" has a very low potential for the provision of habitat services and is limited to 0.66% in terms of area. The ecosystem type class "suboceanic, moderately dry to fresh, moderately nutrient-rich spruce forest" occurs on 9.13% of the area of the Kellerwald National Park and has a low potential for the provision of habitat services. This ecosystem type class is particularly represented at the national park boundaries.

With a percentage occurrence of 72%, the "high potential" category is dominant. In this category, the ecosystem type class "suboceanic, moderately dry to fresh, moderately nutrient-rich fir-beech forest" is the most common ecosystem type class with a total percentage share of 70.37%. The "very high potential" category has a low percentage share of 1.73% in the study area. These areas are mainly located in the far north and east of the national park.

For net phytomass production, the categories "no significant potential", "very low potential" and "low potential" are not represented in the study area. The ecosystem type classes "suboceanic, moderately dry to fresh, poor larch forest", "suboceanic, moist, nutrient-rich black alder floodplain forest" and "suboceanic, dry, nutrient-rich, carbonate-containing sessile oak rocky dry forest" with medium phytomass potential only account for a very small proportion of the area, with a percentage share of less than 2%. These areas occur only sporadically throughout the Kellerwald National Park. With a percentage share of 98%, the "high potential" category covers almost the entire area of the national park. With a share of 70%, the ecosystem type class "suboceanic, moderately dry to fresh, moderately nutrient-rich fir-beech forest" is the most common class, followed by the ecosystem type class "suboceanic, moderately dry to fresh, moderately nutrient-rich spruce forest" (16.6%). The most productive ecosystem type class "suboceanic, moderately dry to fresh, nutrient-rich beech forest" (0.3%) occurs in the centre of the national park, among other places.

The carbon storage capacity of the forest soils in the Kellerwald National Park is distributed across the categories from "very low potential" to "high potential". The ecosystem type class "suboceanic, moderately dry to fresh, poor larch forest" has a percentage share of 0.66% in the Kellerwald National Park and is the only class with a very low potential for carbon sequestration in forest soils. The "low potential" category is the second most common category for carbon storage in the Kellerwald-Edersee National Park. In this category, the ecosystem type class "suboceanic, moderately dry to fresh, moderately nutrient-rich beech forest" and the ecosystem type class "suboceanic, moderately dry to fresh, moderately nutrient-rich spruce forest" are represented. The "medium potential" category is the dominant category in terms of area. The ecosystem type class "suboceanic, moderately dry to fresh, moderately nutrient-rich fir-beech forest" alone accounts for 70% of the national park. A high and very high potential for carbon storage is rather rare in the Kellerwald National Park, with a percentage occurrence of 2.3%. Ecosystem type classes with high potential are primarily the "suboceanic, moderately dry to fresh, nutrient-rich beech forests". The areas with high potential belong to the ecosystem type class "suboceanic, moderately dry to fresh, nutrient-rich beech forest".

The comparison of the results of the assessment of ecosystem services at national and local level shows that the rule-based methodology can be applied at all scales. Only the resolution or scale level of the basic data is decisive for the regional resolution of the results and ultimately for their plausibility.

This shows another advantage of the rule-based assessment method for ecosystem services: The predominantly high assessment of, for example, habitat service in the Kellerwald National Park is objectively substantiated on the basis of quantifiable indicators. Other assessment methods, in particular a verbal-argumentative expert assessment, could only have led to a relative differentiation from low to very high in the national park area. However, this would not have adequately reflected the absolute national importance of the national park.

5.3   Ecosystem Services Under Climate Change in Saxony's Forests

Schlutow and Gemballa (n.d.) carried out a project on the adaptation of the climate structure and the leading forest communities to climate change in the Free State of Saxony.

The results for the periods 1991–2020 and 2041–2070 according to the climate projection RCP8.5 (p1) can be interpreted as follows: The comparison of the assessment of habitat services between the periods 1991–2000 and 2041–2070 according to the climate projection RCP8.5 (p1) makes it clear that the values in the North Saxon Lowlands will decrease. The reason for this is the significant lengthening of the vegetation period with a simultaneous decrease in the climatic water balance in the vegetation months. In future, forest communities from south-east Europe will have the best chances of survival here. However, as these cannot migrate across the Alps under their own steam, artificial afforestation initially requires the decades-long abandonment of the typically associated ground vegetation and the corresponding fauna, which would be necessary to maintain resilience to pest calamities.

In the hilly and montane areas of the southern part of Saxony, on the other hand, habitat services are increasing. This is due to the fact that the currently still more or less pure coniferous forest stands in the higher mountainous areas are to be replaced by deciduous and mixed forest communities in the future. However, deciduous and mixed forests are more biodiverse, resilient and stable than purely coniferous forests.

The biomass productivity of forest trees will not change significantly in the future, as the nutrient conditions of the soil (after a strong wave of acidification and eutrophication in the second half of the twentieth century) will not change significantly. The recommended leading wood communities for natural forest regeneration in Saxony are adapted to rising temperatures, longer vegetation periods and a decreasing climatic water balance in the vegetation period and can therefore guarantee an almost constant wood production (with more stable vitality and higher resilience).

The binding of carbon in the organic soil layer will decrease almost everywhere in Saxony in the future. The rising mean daily temperature leads to faster mineralisation of the dead organic matter and, in combination with this, to faster emissions of CO_2.

As the example shows, the rule-based assessment methodology for ecosystem services has proven its worth in the optimisation of ecological design measures. In the planning process, a rule-based comparison of possible target vegetation types can be used to identify the one that has the best target state for one, several or all relevant ecosystem services.

## 5.4	Ecosystem Services Under the Influence of Climate Change and Air Pollution at the LTER Site

The results allow the following conclusions to be drawn: The habitat service has been drastically reduced by the conversion to an unstructured, mixed beech-pine forest that is not typical of the site. However, the resulting effect of increased biomass production did not materialise. Carbon sequestration has decreased in recent decades due to rising autumn temperatures and will continue to decrease. The higher temperatures accelerate the mineralisation of litter and thus the release of CO_2 into the atmosphere. The current artificial beech-pine forest should not be converted back into a beech forest in the course of near-natural forest conversion due to the increasing atmospheric dryness. A climatic water balance below -22 mm a^{-1} restricts the vitality of beech to such an extent that it is no longer able to compete with other tree species that are more resistant to atmospheric dryness, such as lime, hornbeam and above all English oak. In the past, high nitrogen inputs have not only led to eutrophication, but also to soil acidification. Nitrogen accumulation due to nitrogenous air pollutants can hardly ever be completely reversed. Before the wave of nitrogen inputs in the second half of the twentieth century, there was a nitrogen deficiency in Germany's soils. In order to restore the previous nitrogen status, the uptake rate into the biomass would have to be higher than the input rate plus the mineralisation rate. This does not appear to be in sight at present.

At the same time, however, the acid-base state typical of the Cambisols will be restored from eutrophic sandy deposits if deep-rooted oaks are introduced with the forest reorganisation in the future, which transport basic cations from deeper soil layers into the leaves and thus shift the bases into the topsoil after the litter fall. Although today's pines are also deep-rooted, they also contribute to the acidification of the topsoil with their needle litter.

A planned near-natural forest reorganisation should therefore focus on deep-rooted deciduous tree species that can withstand the expected rising temperatures and increasing drought. One such sustainable forest community suitable for the site could be the spurge and downy oak community (*Euphorbio angulatae-Quercetum pubescentis*) with a mixture of downy oak (*Quercus pubescens*), oak (*Quercus cerris*), field maple (*Acer campestre*) and many shrub species (*Cornus mas, Cornus sanguinea, Ligustrum vulgare, Rosa spec, Viburnum lantana, Crataegus spec., Euonymus verrucosa*) in the undergrowth.

As the example shows, the development of the ecological service of an ecosystem depends not only on its current state, but also on its past development. Thus, current ecosystem services can often only be explained if the development history is also known and the corresponding effects on the current state can be interpreted. It therefore makes sense to apply the same rules and quantitative indicators for the rule-based assessment of ecosystem services to specific time slices of a time series of influencing factors, such as substance inputs and/or climate parameters, and to document their changes.

5.5   Ecosystem Services of Open Land Habitats

The application of the rules for assessing habitat service means that some biotope types are rated just as highly as forests. These are in particular unused or extensively utilised biotopes, especially on extremely wet or extremely dry sites.

Net biomass production in non-forest areas is generally significantly lower than in forest areas. Nevertheless, floodplain and alluvial meadows as well as wetland and marsh meadows achieve an overall rating comparable to that of forests.

Applying the rules for assessing carbon storage capacity leads to a result that was to be expected for the relevant open land biotopes. Wetlands and moors have particularly high carbon storage capacities. However, heathlands also accumulate organically bound carbon due to the heavy and thus slow decomposition of the woody heathland biomass and its high proportion of residues after grazing.

The rules and quantitative indicators established for the assessment of ecosystem services can therefore, in principle, also be applied to open land areas without further ado.

References

Bösch B (2001) Neue Bonitierungs- und Zuwachshilfen, Schriftenreihe Freiburger Forstliche Forschung, Wissenstransfer in Praxis und Gesellschaft, vol 18. FVA-Forschungstage. https://www.waldwissen.net/technik/inventur/fva_schaetzhilfen/fva_schaetzhilfen.pdf

Dirnböck T, Djukic I, Kitzler B, Kobler J, Mol-Dijkstra JP, Posch M, Reinds GJ, Schlutow A, Starlinger F, Wamelink GWW (2017) Climate and air pollution impacts on habitat suitability of Austrian forest ecosystems. PLoS One 12(9):e0184194. https://doi.org/10.1371/journal.pone.0184194

MAES (2016) Mapping and assessment of ecosystems and their services. http://biodiversity.europa.eu/maes. Accessed 25 July 2019

Schlutow A, Gemballa R (n.d.) Ableitung von klimawandelangepassten Leitwaldgesellschaften für die Wälder im Freistaat Sachsen [Derivation of climate change-adapted indicator forest communities for the forests in the Free State of Saxony]. Staatsbetrieb Sachsenforst (ed.). Pirna

Schlutow A, Scheuschner T (2023) Determination of critical loads for eutrophying and acidifying air pollutant inputs for the protection of near-natural ecosystems in Germany. Atmos 14(2):383. https://doi.org/10.3390/atmos14020383

Schlutow A, Schröder W (2021) Rule-based classification and mapping of ecosystem services with data on the integrity of forest ecosystems. Environ Sci Eur 33:50. https://doi.org/10.1186/s12302-021-00481-3

Schlutow A, Schröder W, Scheuschner T (2021) Assessing the relevance of atmospheric heavy metal deposition with regard to ecosystem integrity and human health in Germany. Environ Sci Eur 33(1). https://doi.org/10.1186/s12302-020-00391-w

Schlutow A, Kraft P, Scheuschner T, Schlutow M, Schröder W (2024) Bioindication for Ecosystem Regeneration towards Natural conditions—the BERN data base and BERN model. Environ Sci Eur 36(1). https://doi.org/10.1186/s12302-023-00826-0. BERN database (open source) under. https://github.com/bern-model/BERN

Wellbrock N, Bolte A, Flessa H (Hrsg.) (2017) Dynamik und räumliche Muster forstlicher Standorte in Deutschland—Ergebnisse der Bodenzustandserhebung im Wald. [Dynamics and spatial patterns of forest sites in Germany—results of the soil condition survey in forests]. https://www.thuenen.de/media/institute/wo/Waldmonitoring/bze

Chapter 6
Recommendations

Abstract The methodology just is applicable to larger-scale regional and local scales as to single sites.

The classification of ecosystem type classes according to the proposed rules allows the transfer of the methodology to territories outside Germany, especially to Central Europe. The basis for the designation of ecosystem classes would have to be a soil map, a climate map and a vegetation type map respectively.

Keywords Method applicability · European map

In general, the ecosystem services of pristine ecosystem type classes can be compared with ecosystem services under current and expected future environmental conditions, such as atmospheric nitrogen deposition and climate change in particular (Gaudio et al. 2015; Kwon et al. 2021; Matyssek et al. 2012; Schaub et al. 2021), from which the need and type and scope of measures to restore the highest possible performance can be derived. This requires a rule-based assessment of ecosystem services using quantitative indicators so that the process is comprehensible and transparent and the results are reproducible. Only such a transparent, quantitative methodology for assessing ecosystem services makes it possible to objectively determine the current deviation of ecosystem services from the original state and thus to determine a realistic need for action to restore the ecosystem service potential. The following applies: the smaller the deviation of the currently identifiable (possibly anthropogenically impaired) ecosystem services from the potential of the ecosystem, the higher the functional capacity of the ecosystem is to be assessed. The first prerequisite for this was an assessment of the service of the ecosystem type class in its original state, as illustrated here using the example of three ecosystem services.

The categorisation of ecosystem type classes according to the proposed rules (Sect. 3.1.1) makes it possible to transfer the methodology to areas outside Germany,

© The Author(s), under exclusive license to Springer Nature
Switzerland AG 2024
A. Schlutow, W. Schröder, *Climate Change and Atmospheric Deposition as Drivers of Forest Ecosystem Integrity and Services*, SpringerBriefs in Environmental Science, https://doi.org/10.1007/978-3-031-67103-6_6

in particular to Central Europe. The basis for the designation of ecosystem classes should be a soil map, a climate map or a vegetation type map.

The Eurosoil soil mapping (European Soil Data Centre (ESDAC) 2020; Eurosoil 1999; Panagos et al. 2020), the European climate grid (DWD 2019), the mapping of Europe's natural vegetation (EuroVegMap 2000) and the BERN database (Schlutow et al. 2024) are available for the whole of Europe. The methodology can be applied to larger regional and local scales as well as to individual locations. The basis for the designation of ecosystem classes should be a soil map, a climate map or a vegetation type map. For the ecosystem classes, the rule-based point values of the ecosystem services could be determined in accordance with Sect. 2.

References

DWD (Deutscher Wetterdienst) (2019) Mittlere Tagesmitteltemperatur, mittlere Niederschlagsmengen aus dem Zeitraum 2001–2014. Rasterdatei [Mean daily temperature, mean precipitation from the period 2001–2014. Raster file]. https://opendata.dwd.de/climate_ environment/CDC/grids_europe/. Assessed on 23.02.2023

European Soil Data Centre (ESDAC) (2020) JRC support to the European Joint Programme for soil (EJP SOIL). In: Panagos P, Jones A, Van Liedekerke M, Orgiazzi A, Lugato E, Montanarella L (eds) 2020EUR 30450ENDatasets, technical advice and scientific guidance. https://esdac.jrc. ec.europa.eu/public_path/u890/_Pubs/EUR30450.pdf

Eurosoil (1999) Metadata: soil geographical data base of Europe, vol 3.2.8.0. Joint Research Centre, Ispra

EuroVegMap (2000) Map of the natural vegetation of Europe: scale 1:2,500,000, Bundesamt für Naturschutz (German Federal Agency for Nature conservation), Bonn, 2000. http://www.flo-raweb.de/download/eurovegmap/natural_vegetation_toplevelunits.xls

Gaudio N, Belyazid S, Gendre X, Mansat A, Nicolas M, Rizzetto S (2015) Combined effect of atmospheric nitrogen deposition and climate change on temperate forest soil biogeochemistry: a modeling approach. Ecol Model 306:24–34

Kwon T., Shibata H., Kepfer-Rojas S., Schmidt I.K., Larsen K.S., Beier C., Berg B., Verheyen K., Lamarque J.-F., Hagedorn F., Eisenhauer N., Djukic I., (Tea Composition Network) (2021): Effects of climate and atmospheric nitrogen deposition on early to mid-term stage litter decomposition across biomes. Front For Glob Chang 4:678480. doi: https://doi.org/10.3389/ffgc.2021.678480

Matyssek R, Wieser G, Calfapietra C, De Vries W, Dizengremel P, Ernst D, Jolivet Y, Mikkelsen TN, Mohren GMJ, Le Thiec D, Tuovinen J-P, Weatherall A, Paoletti E (2012) Forests under climate change and air pollution: gaps in understanding and future directions for research. Environ Pollut 160:57–65. https://doi.org/10.1016/j.envpol.2011.07.007

Panagos P, Jones A, Van Liedekerke M, Orgiazzi A, Lugato E, Montanarella L (2020) JRC support to the European joint Programme for soil (EJP SOIL), EUR 30450 EN. Publications Office of the European Union, Luxembourg, p JRC1222. https://doi.org/10.2760/74273

Schaub M, Vesterdal L, De Vos B, Ukonmaanaho L, Fleck S, Schwärzel K, Ferretti M (2021) Forest Monitoring to assess forest functioning under air pollution and climate change. FORECOMON 2021—The 9th Forest Ecosystem Monitoring Conference, 7–9 June 2021, Birmensdorf, Switzerland. Proceedings. Birmensdorf, Swiss Federal Research Institute WSL, pp 1–108. https://doi.org/10.16904/envidat.225. https://www.icp-forests.org/pdf/SC2021_pro-ceedings.pdf

Schlutow A, Kraft P, Scheuschner T, Schlutow M, Schröder W (2024) Bioindication for ecosystem regeneration towards natural conditions—the BERN data base and BERN model. Environ Sci Eur 36(1). https://doi.org/10.1186/s12302-023-00826-0. BERN database (open source) under. https://github.com/bern-model/BERN